Ethical Conflicts In Information and Computer Science, Technology, and Business

Books and Training Products From QED

DATABASE

Data Analysis: The Key to Data Base Design
The Data Dictionary: Concepts and Uses
DB2: The Complete Guide to Implementation and Use
Logical Data Base Design
DB2 Design Review Guidelines
DB2: Maximizing Performance of Online Production Systems
Entity-Relationship Approach to Logical Data Base Design
How to Use ORACLE SQL*PLUS
ORACLE: Building High Performance Online Systems
Embedded SQL for DB2: Application Design and Programming
SQL for dBASE IV
Introduction to Data and Activity Analysis
ORACLE Design Review Guidelines
Using DB2 to Build Decision Support Systems
How to Use SQL for DB2

SYSTEMS ENGINEERING

Handbook of Screen Format Design
Managing Projects: Selecting and Using PC-Based Project Management Systems
The Complete Guide to Software Testing
A User's Guide for Defining Software Requirements
A Structured Approach to Systems Testing
Practical Applications of Expert Systems
Expert Systems Development: Building PC-Based Applications
Storyboard Prototyping: A New Approach to User Requirements Analysis
The Software Factory: Managing Software Development and Maintenance
Data Architecture: The Information Paradigm
Advanced Topics in Information Engineering

MANAGEMENT

CASE: The Potential and the Pitfalls
Strategic and Operational Planning for Information Services
The State of the Art in Decision Support Systems
The Management Handbook for Information Center and End-User Computing
Disaster Recovery: Contingency Planning and Program Analysis

MANAGEMENT (cont'd)

Winning the Change Game
Information Systems Planning for Competitive Advantage
Critical Issues in Information Processing Management and Technology
Developing the World Class Information Systems Organization
The Technical Instructor's Handbook: From Techie to Teacher
Collision: Theory vs. Reality in Expert System
How to Automate Your Computer Center: Achieving Unattended Operations
Ethical Conflicts in Information and Computer Science, Technology, and Business

DATA COMMUNICATIONS

Data Communications: Concepts and Solutions
Designing and Implementing Ethernet Networks
Network Concepts and Architectures
Open Systems: The Guide to OSI and its Implementation
VAX/VMS: Mastering DCL Commands and Utilities

PROGRAMMING

VSAM Techniques: Systems Concepts and Programming Procedures
How to Use CICS to Create On-Line Applications: Methods and Solutions
DOS/VSE/SP Guide for Systems Programming: Concepts, Programs, Macros, Subroutines
Systems Programmer's Problem Solver
VSAM: Guide to Optimization and Design
MVS/TSO: Mastering CLISTS
MVS/TSO: Mastering Native Mode and ISPF
VAX/VMS: Mastering DCL Commands and Utilities

SELF-PACED TRAINING

SQL as a Second Language
Building Online Production Systems with DB2 (Video)
Introduction to UNIX (CBT)
Building Production Applications with ORACLE (Video)

For Additional Information or a Free Catalog contact

QED INFORMATION SCIENCES, INC. • P. O. Box 82-181 • Wellesley, MA 02181
Telephone: 800-343-4848 or 617-237-5656

Ethical Conflicts In Information and Computer Science, Technology, and Business

Prepared by:

Donn B. Parker
Susan Swope
Dr. Bruce N. Baker

QED Information Sciences, Inc.
Wellesley, Massachusetts

© 1990 by QED Information Sciences, Inc.
P.O. Box 82-181
Wellesley, MA 02181

Library of Congress Catalog Number: 89-38890
International Standard Book Number: 0-89435-313-6

Printed in the United States of America
90 91 92 10 9 8 7 6 5 4 3 2

Library of Congresss Cataloging-in-Publication Data

Parker, Donn B.
  Ethical conflicts in information and computer science, technology, and
business/Donn B. Parker, Susan Swope, Bruce N. Baker.
   p. cm.
ISBN: 0-89435-313-6
1. Business ethics. 2. Information science—Moral and ethical aspects.
3. Technology—Moral and ethical aspects. 4. Computer crimes.
5. Responsibility. I. Swope, Susan. II. Baker, Bruce N. III. Title
HF5387.P367 1989
174'.9004—dc20

Contents

III.   PROPERTY OWNERSHIP, ATTRIBUTION, PIRACY, PLAGIARISM, COPYRIGHTS, AND TRADE SECRETS 63

I
Introduction

We learn our basic ethical principles in the formative years of childhood. As we mature, however, ethics take on new meaning and importance, especially as we begin a professional career. In our chosen field, we are expected to augment and interpret the ethical values we learned earlier. This study is not intended to change people's ethical values, but rather to assist users of computers and data communications in clarifying and applying their ethical values as they encounter new, complex situations where it may not be obvious how ethical values may apply or where the appropriate application of one of these values may conflict with other ethical values.

Advancements in computer and data communications technology have resulted in the need to reevaluate the application of ethical principles and establish new agreements on ethical practices. The application of ethics in information science, technology, and business is more difficult than in other disciplines for several reasons. First, computers and data communications alter relationships among people. Data communications take place without personal contact, without the visual and aural senses to help convey meaning. Moreover, the paperless society, in which information is transmitted at electronic speeds, functions side by side with the paper-based society, where information is shared at a snail's pace. Conveying one's intentions in a letter, which can take days to reach the recipient, is very different from instantaneous electronic transmission because of how quickly the recipient may act on them.

Communication occurs so quickly that one may not have time to consider the implications of the information before it has been sent and received.

Second, information in electronic, magnetic, and optical form is far more fragile than information on paper. Computers and data communications systems provide for high-speed, low-cost processing, communication, copying, and printing of intangible intellectual property. This capability introduces new factors in decisions about proprietary rights, residual rights, plagiarism, piracy, eavesdropping, and violation of privacy. Negative events happen so easily, sometimes without the initiators' even considering the consequences, that ethical issues are intensified. Freedom of expression is greatly leveraged and magnified to the extent that far more good may be done with the creation, use, and dissemination of information. Yet it follows that the consequences of unethical acts are equally magnified.

Among computer and data communications users, efforts to protect the integrity, confidentiality, and availability of information inevitably conflict with the apparent benefits of sharing information. The principle of "need to know," long enforced in national defense, is now prevalent in academic, research, and business environments; that principle sometimes clashes with the important value of free access to information.

Finally, business transactions rely on handwritten signatures, yet nearly all electronic transactions take place without any signatures. As automated means for transmitting legally recognized signatures become available, ethical norms will need to change.

Unlike the computer field, some sciences and professions have had hundreds of years to develop the ethical concepts that form the basis for dealing with new issues. They nonetheless continue to wrestle with new and troublesome ethical problems raised by technological advances. In contrast, computer science and technology have been in existence for only 30 years. The need for ethical standards is equally as critical in computer science, technology, and business as it is in other fields. Given the problems in the ancient sciences, it is little wonder that serious problems arise in developing ethical concepts and practices in such a comparatively new field of knowledge.

In addition, ethical issues in the computer field may be more separable and public than in medicine and law, which are well-defined professions with limited membership. In those professions, decisions can often be made out of public view. On the other hand, many more people in widely diverse situations decide the ethicality of computer issues. The general principles discussed in this study will often affect average citizens as well as the highly specialized professionals.

Some of the computer-specific ethical issues that have arisen as a result of the development of computers include:

- Repositories and processors of information. Unauthorized use of otherwise unused computer services or of information stored in computers raises questions of appropriateness, fairness, invasion of privacy, and the public's right to know (freedom of information).

- Producers of new forms and types of assets. Computer programs, for example, are entirely new types of assets that may not be subject to the same concepts of ownership as other assets.

- Instruments of acts. To what degree must providers of computer services and users of computers, data, and programs be responsible for the integrity and appropriateness of their computer output?

- Symbols of intimidation and deception. The anthropomorphic view of computers as thinking machines, infallible absolute-truth producers, that yet are subject to blame should be carefully considered.

These roles of computers have been carefully documented at SRI International in continuing studies of over 2,500 reported cases of intentionally caused losses associated with computers since 1958. This study of computer abuse, combined with a concern about the root causes of unauthorized acts, led to the 1977 study (Ethical Conflicts in Computer Science and Technology, Donn B. Parker, AFIPS Press, 1981), as well as this present study on ethics.

Focus of the Study

Advancing technology has changed the relationships between individuals and organizations as well. This study considers only the ethicality of individual actions, although organizations, too, can act ethically or unethically. The ethical decisions of members of an organization cannot always be easily separated from those of the entire group. Yet organizational actions can always be traced to individual people — these are the individuals held accountable for their actions in the scenarios presented in this book.

Social movements also can significantly affect ethical decisions. For example, if society were to conclude that the Strategic Defense Initiative (SDI) is an acceptable deterrent to nuclear holocaust and could be effectively implemented, then individual scientists who disagree and refuse to participate might be judged to be acting unethically, even though they are adhering to their personal standards. On the other hand, if politicians and military experts cannot agree on the advisability of SDI, then an individual scientist's refusal to support SDI is more likely to be viewed as ethical. The underlying ethical principles must be identified in such cases to conclude whether individual actions are appropriate, independent of social movements. Again, the focus of this study is on personal accountability to ethical principles and not necessarily on organizational or social acceptability of what is ethical.

In 1977, during the first computer ethics study, computers were not nearly so numerous nor so networked together as they are today. Individuals functioned strictly as computer professionals, computer scientists, and computer teachers serving those lay persons who used the output of computers. Now, because of the widespread use of computers, distinguishing between specialists who work only with computers and those who use computers to perform specialized tasks in other disciplines lacks significance.

Computer education now begins in grade school; it is no longer a restricted technical specialty, learned only as part of on-the-job training or in an engineering or mathematics graduate program. Computers have become as commonplace as telephones. The related ethical issues have thus become more democratically defined. More people have more to say about computer ethics simply because so many more people are computer-literate. On the other hand, the diffuseness of the impacts and the wide distribution of the technology mean that recognizing impacts, let alone solving an ethical dilemma, is much more difficult.

Written codes of ethics may no longer be needed to deal with computer and telecommunications technology. Ethical principles applied to millions of computer users effectively become the equivalent of common law. Informed consent may be posed as a panacea for guidance on ethical decisions. If everyone affected by a decision is consulted before an act is committed, all parties and their well-being can be adequately considered; each person can take a position and influence the action. Clearly, life is not so simple. As the scenarios show, computer personnel often cannot obtain such universal consent, primarily because of a lack of time or resources or ignorance of who may be affected. The scenarios assume that efforts to obtain informed consent have already been reasonably exhausted without reaching an ethical decision. What remain are the difficult choices.

Individuals could apply the higher ethical principle to resolve the ethicality of an act. That is, a more compelling factor may exist to enforce a decision that might otherwise be considered unethical. For example, a person who, to avoid being fired, simulates an active session with a mainframe computer from a microcomputer terminal located at home (as though busy at work) while rushing a sick child to the hospital might be satisfying a higher ethical principle. Such extenuating circumstances are not allowed for consideration in the study scenarios, however, unless explicitly stated.

Additional principles that are intrinsically part of evaluating the scenarios are the Kantian universality rule and the Descartian observation that, respectively, an act that is not right for everyone is not right for anyone and a sufficient change in degree produces a change in kind. Thus, use of an employer's computer resources without authorization is not justified on the basis that the resources would otherwise go unused. If everybody did that, no resources would be available for authorized work. Taking an increasing amount of computer resources for unauthorized personal use at some point changes from immaterial to material unethical behavior.

Finally, despite the careful analysis by all workshop participants, one challenging and difficult question is unfortunately still left unanswered: Does changing technological sophistication raise new ethical issues, or do the ethics questions remain constant?

Background

In 1970, the American Federation of Information Processing Societies (AFIPS) sponsored a roundtable on professionalism in the computer field chaired by the Honorable Willard Wirtz, former Secretary of Labor. The participants recognized the need for formalized ethical standards in the computer field. Since then, constituent society members of AFIPS, such as the Association for Computing Machinery (ACM) and the Data Processing Management Association, have developed codes of ethics but have done little to consummate them by establishing sanctions, enforcing them, or testing their applicability. The Institute for Certification of Computer Professionals (ICCP) also has a code of ethics associated with the Certificate in Data Processing program. Several applicants have been refused certification for failure to meet certain requirements of the code, but to our knowledge the code has never been used to sanction anyone holding a certificate.

More important, the codes have never been interpreted, primarily because they have so seldom been applied. Therefore, the ethicality of any but the most obvious situations covered by the codes is unknown. The legal system, however, is being used to settle an increasing number of issues in the computer science, technological, and business community.

Most states have computer crime statutes, and two federal laws have been enacted — the Computer Fraud and Abuse Act of 1986 (18 U.S.C., Ch. 47, Sect. 2101-2103, Art. 1030) and the Electronic Communications Privacy Act of 1986 (18 U.S.C., Ch. 119, Sect. 2501). Legality, however, is not the same as professional ethics. Leaders in computer and communications technology, science, and business need to recognize the ethical conflicts that professionals, computer scientists, and employees face, while management needs to support those who responsibly follow ethical principles. However, codes of ethics must not place professionals in impossible moral and economic situations. Scientific, professional, and trade associations must coordinate their efforts to identify the ethical issues, to determine the ethicality of actions that may be taken in situations when those issues arise, and to promulgate the resulting ethical values among businesses, institutions, and government.

Developing a New Approach

Developing a theoretical code of ethics before the issues have been fully identified and documented, and before a consensus has been reached on what constitutes ethical or unethical action, seems a sterile and unproductive approach. An empirical, pragmatic approach based on identified issues is necessary. One means of carrying out such an approach is by using ethical issue scenarios. Fortunately, at the time this idea was generated in late 1975, a possible source of financial support became available in a new program called Ethics and Values in Science and Technology (EVIST), in the Office of Science and Society, Science Education Directorate of the National Science Foundation.

In response to an SRI International proposal in February 1976, the National Science Foundation awarded a grant to conduct a computer science and technology ethics workshop. The project was completed in 1977 and the findings published in Ethical Conflicts in Computer Science and Technology, by Donn B. Parker (AFIPS Press, 1981). SRI developed a new, experimental approach to discussing ethical issues, scenario analysis.

Ten years later in 1987, Kathleen Gagne at AFIPS Press suggested to Donn B. Parker at SRI International that it might be timely to repeat the original 1977 study considering the proliferation, advancement, and changes in the use of computers. Subsequently, Mr. Parker obtained a grant for SRI from the U.S. National Science Foundation, Directorate for Biological, Behavioral and Social Sciences, Ethics and Values Studies, with support from the Directorate of Computer Sciences, for The Second Computer Science and Technology Ethics Workshop.

The Use of Scenario Analysis Methodology

The goal of this third study (the first was the Willard Wirtz roundtable on professionalism in 1970 and the second was the 1977 NSF grant study at SRI) was to further develop the concepts of ethical and unethical practices that are unique or prevalent in the fields of computer science and technology while also encompassing the business community. As part of the study, workshop participants were to consider a set of scenarios depicting ethical problems and suggest general principles concerning ethical and unethical practices in the computer field where ethical decisions are crucial.

The final results were to be used to suggest more explicit ethical principles for computer users and to provide text material for courses dealing with social issues in colleges, universities, and trade schools for students entering the computer field. The published results were to update the ethical and unethical practices in the computer field described in the 1977 study. The models for the proposed published work are the first book and the book Ethical Problems in Engineering, by Alger, Christensen, and Olmsted (John Wiley & Sons, 1965).

Participants were selected on the basis of their known interests in ethics in the computer field. They represented a wide range of interests, including both business professionals and those with backgrounds in the computer field, ethical philosophers, and lawyers — all were considered essential participants in such a group. Four of the invitees had participated in the first NSF workshop. Two ethical philosophers, Dr. Deborah Johnson, Associate Professor of Philosophy at Rensselaer Polytechnic Institute, and Dr. Susan Leigh Star of Tremont Research Institute, assisted SRI by reviewing scenarios, developing scenarios, participating in the workshop, and otherwise providing expert guidance.

ETHICS WORKSHOP PARTICIPANTS — JUNE 11–12, 1987
*Unable to attend the workshop

Robert P. Abbott*
President
EDP Audit Control
Oakland,. CA 94603

Robert H. Anderson
Director
Institute for Research on Interactive Systems
The RAND Corporation
Santa Monica, CA

J. J. Bloombecker
The National Center for Computer
Crime Data
Los Angeles, CA

David Brandin*
President
Strategic Technologies
Los Altos Hills, CA

David Burnham
Freelance Writer
Washington, DC

John M. Carroll
Professor
Department of Computer Science
The University of Western Ontario

Gary Chapman
Executive Director
Computer Professionals for Social Responsibility
Palo Alto, CA

Scott Cook
Assistant Professor of Philosophy
Department of Philosophy
San Jose State University
San Jose, CA

Richard T. De George
Professor
Department of Philosophy
University of Kansas
Lawrence, KS

Frank Dubinskas*
Associate for Case Development
Harvard Business School
Cambridge, MA

Albert Flores
Professor of Philosophy
California State University Fullerton
Fullerton, CA

Bernard A. Galler
Professor
Electrical Engineering and Computer Sciences
Ann Arbor, MI

Bruce Gilchrist
Senior Advisor for Information Strategy
Columbia University
New York, NY

Calvin Gotlieb
Department of Computer Science
University of Toronto
Toronto, Canada

Edwin B. Heinlein
President/Owner
Heinlein Associates, Inc.
San Rafael, CA

Deborah Johnson
Associate Professor of Philosophy
Department of Science and Technology Studies
Rensselaer Polytechnic Institute
Troy, NY

David Kadlecek
Nanometrics, Inc.
Sunnyvale, CA

Ronald G. Keelan*
Program Director
Data Security Programs
EBM Corporation
Purchase, NY

John McLeod, P. E.
Chairman, Ethics Committee
The Society for Computer Simulation
La Jolla, CA

Jeffrey A. Meldman
Senior Lecturer in Management and
    Associate Dean for Student Affairs
Massachusetts Institute of Technology
Cambridge, MA

Jim Moor
Professor of Philosophy
Philosophy Department
Dartmouth College
Hanover, NH

Arthur L. Norberg*
Director
Charles Babbage Institute
Minneapolis, MN

Peter Neumann
Staff Scientist
Computer Science Laboratory
SRI International
Menlo Park, CA

Susan H. Nycum
Baker and McKenzie
Palo Alto, CA

Dr. Andrew Oldenquist
Professor of Philosophy
Mershon Center Associate
Department of Philosophy
Ohio State University
Columbus, OH

Mark Pastin*
Director, Center for Ethics
Professor of Management, Center for Ethics
Arizona State University
Tempe, AZ

Danielle Pouliot
Criminologist/Consultant
DMR Group, Inc.
Montreal, Quebec, Canada

Robert Riser
East Tennessee State University
Johnson City, TN

Mr. Marc Rotenberg
Member
Computer Professionals for Social Responsibility
Palo Alto, CA

John W. Snapper
Associate Professor of Philosophy
Illinois Institute of Technology
Chicago, IL

Susan Leigh Star
Director, Methods Research
Tremont Research Institute
San Francisco, CA

Rein Turn*
Professor of Computer Science
California State University, Northridge
Northridge, CA

Alan F. Westin
Professor of Public Law and Government
Columbia University
Teaneck, NJ

Joel Yudkin
Member
Computer Professionals for Social Responsibility
Menlo Park, CA

Initially, SRI developed 24 new scenarios and selected 17 scenarios from the 1977 study that were the most controversial or where opinions were expected to have changed. The scenarios are short, one-page, simple stories based in part on SRI International's computer abuse research. The stories were limited to individuals engaging in acts or confronted with acts that might be construed as unethical. Issues of a global nature — for example, issues about computers being used as instruments of war or for suppression of human rights — that would require joint decisions of many people in organizational situations were avoided. Only personal issues facing individuals in computer science, technology, and business were covered. The following model was used:

Subjects	Environments	Objects
Consumer/public	Academic	Physical property
Computer users	Corporate business	Intellectual property
Students	Corporate institutions	Financial assets
Teachers	Government	Use of services
Scientists	Independent	
Consultants		
Analysts		
Programmers		
Computer operators		
Manufacturer representatives		
Managers		

The scenarios were designed to raise questions of unethicality rather than ethicality. A minimum of details was provided. Descriptions of acts were made as objective as possible, and superlative and negative adjectives were avoided wherever practical.

Copies of the scenarios were sent to all participants before the workshop. Participants were asked to vote on each actor and act, indicating whether each was unethical, not unethical, or not an ethics issue. The participants were not asked to indicate whether they thought the acts were ethical or in any way exemplary. Because the project was designed to consider only those acts having ethical connotations, the not-an-ethics-issue vote was included to account for the lack of an ethical issue or a clear violation of the law.

The differences among such words as illegal, unethical, tortuous, immoral, unconventional, incompetent, and unwise blurred when participants were forced to limit their votes on the scenarios to unethical, not unethical, or no ethics issue. In the context of the ethics workshop, "unethical" means not conforming to an appropriate personal standard of conduct, "not unethical" means not violating an appropriate personal standard of conduct, and "no ethics issue" means no appropriate personal standard of conduct was involved or the action was clearly illegal. These definitions are the essence Of ethicality. Yet they caused extensive and sometimes heated discussions because participants had differing personal standards, partly as a result of differences in experiences and backgrounds.

Participants were then asked to write their opinions about each actor and act that supported their votes, indicating any problems in reaching a decision, and to suggest normative (later changed to general) principles that related to the scenario. A copy of the response form is shown in the accompanying figure. Participants were told that the scenarios must be considered absolute. They were not subject to modification in their deliberations, nor should any more conditions than were explicitly stated in the scenarios be assumed.

ETHICS SCENARIO ANALYSIS   Return to: Donn B. Parker
                                                                SRI
                                                                Menlo Park, CA 94025

Case No._____ Workshop Participant: _____

A. 1. Party: _____ Act:_____

Unethical _____ Not unethical _____ No ethics issue _____
Relevant Factors: _____

2. Party:_____ Act: _____

Unethical _____ Not unethical _____ No ethics issue _____
Relevant Factors: _____

3. Party:_____ Act: _____

Unethical _____ Not unethical _____ No ethics issue _____
Relevant Factors: _____

B. Ethical Judgments:_____

C. Ethical Judgments:_____

D. 1. Scenario Variations: _____

2. Ethical Judgments: _____

3. Normative Principles: _____

After the results of this first mailing were compiled, a number of new scenarios based on participants' suggestions were added. SRI then prepared the working material for the workshop, which consisted of each scenario, followed by all the opinions and general principles provided by the participants. Duplicates or significantly overlapping items were discarded, and the most representative or most detailed item was used. The workshop was held on June 11 and 12, 1987, at SRI International, Menlo Park, California. Sixty scenarios were provided for consideration, 17 from the 1977 workshop and 43 new ones. Participants decided to drop five of the new scenarios and combine two others. This left 54 scenarios that were discussed and are included in this book. Each day, the participants were divided into four subgroups (lawyers and ethical philosophers were distributed as equally as possible to provide uniform representation). During each half-day session, each subgroup considered a set of related scenarios. This allowed for 15 to 20 minutes of discussion for each scenario by at least one subgroup. Each participant thus covered about one-fourth of the scenarios during four different subgroup discussions.

Each subgroup was asked to generate any new opinions and suggest general principles for all assigned scenarios. At the end of each subgroup meeting, the subgroups were asked to vote again on the actors and acts, as most of them had done previously by mail.

After the subgroups voted on their scenarios, plenary sessions were held at the end of each half-day session. The subgroup spokespersons presented the majority and minority positions of their subgroups; subgroup members could add to the brief presentations. Then other workshop participants who had not participated in the specific subgroup discussions were asked to present their positions. Although time constraints prevented in-depth discussion, the positions presented by others, who were not privy to the particular subgroup discussion, sometimes swayed the votes of the entire group to a conclusion opposite to that held by the subgroup. The votes of the subgroup were not, therefore, viewed as recommendations for others to accept but rather as an opening salvo in a spirited debate. Often, in their deliberations on the scenarios considered at the previous workshop, participants reached different conclusions in 1987 than 1977. At least two factors account for the changes. First, the workshops consisted of two substantially different groups of people, and second, the evolution of the industry over the intervening 10 years. Therefore, it is difficult to compare the voting scores.

Several participants had more to say after the workshop, and their comments are included in the Appendix. (They are not included in the text of the book to preserve as accurate a reporting of workshop discussions as possible. However, they reflect additional careful thought on the part of the workshop participants and are highly recommended.)

Participants believed the plenary sessions were too short to hear the arguments for the various positions, and they felt rushed into having to vote with so little time to consider the issues. Nonetheless, the attendees contended they had accomplished a great deal during the 2-day session and that they had contributed to a significant study. The participants also agreed that their attendance at the workshop benefited them greatly, both personally and professionally. The experience made them much more sensitive to the ethical issues associated with their work.

Categorization of Scenarios

For the 1977 workshop, SRI had not categorized the scenarios by topic, environment, object, or subjects in advance of the workshop. During the mailed response activities, however, Dr. Abbe Moshowitz, a participant from the University of British Columbia, suggested a means of classifying the ethical scenarios. This classification scheme was later modified by a group at the workshop. Two of the classifications were combined subsequent to the workshop, and the resulting six classifications were used in the 1981 book as chapter headings under which the scenarios were grouped. The six classifications were as follows:

I.   Conflict over Obligations and Implicit Contractual Arrangements

II.   Disputed Rights to Products
III. Confidentiality of Sensitive Data and Invasion of Privacy, Personal Morality, and Organizational Loyalty
V.   Responsibility for Computer Applications with Unknown or Controversial Consequences
VI.   Responsibility for Disseminating Complete and Accurate Information to Decision-Makers or the Public

For the 1987 workshop, the same categories were used initially, but SRI received several recommendations that resulted in expanding the categories to ten:

I.   Conflict over Obligations and Implicit Contractual Arrangements
II.   Disputed Rights to Products and to Access to Computers or Computer-Generated Information or Services (Piracy)
III. Confidentiality of Sensitive Data and Invasion of Privacy (Secrecy)
IV.   Personal Morality, Organization Loyalty, and the Public Interest (Whistle Blowing)
V.   Responsibility for Computer Applications with Unknown or Controversial Consequences (Bombs)
VI.   Responsibility for Disseminating Complete and Accurate Information to Decision-Makers or the Public (Disclosure)
VII. Conflicts Among Research Priorities or Between Research and Other Priorities
VIII. Responsible Development and Representation of Expertise
IX.   Business Policy and Professional Guidelines
X.   Setting Technical, Professional, and Public Standards

These categories were maintained during the workshop, but some participants suggested further changes or combinations. After several iterations, SRI selected the following categories to create better balance among the critical issues considered and to remove some of the overlaps:

I.   Professional Standards, Obligations, and Accountability
II.   Property Ownership, Attribution, Piracy, Plagiarism, Copyrights, and Trade Secrets

III.   Confidentiality of Information and Personal Privacy
IV.   Business Practices Including Contracts, Agreements, and
Conflict of Interest
V.   Employer/Employee Relationships

Many of the scenarios could logically fit into several categories, but each has been classified according to the overriding issue facing the primary actors.

Chapter VII summarizes all the principles developed during the workshop. They have been categorized according to whom they apply — general public, professional, employer, and employee — so that readers may more easily reference the principles that affect them. In the 1981 book, principles that did not apply specifically to computer professionals, but were standards of general ethical conduct that everyone should observe, were omitted. Professionals, however, should not forget that they, too, are responsible for adhering to general standards of ethical conduct and that professional standards are in addition to, not in place of, these general standards.

Limitations of the Project and Suggested Interpretations

Because the scenarios are based on composites of actual experience, similarities may exist between a scenario and a real incident. However, it would be totally irresponsible to associate these scenarios with any one person or organization. The scenarios are brief and highly stylized; therefore, they are to be treated as completely fictitious and any similarity to real persons or events as purely coincidental.

Condensing diverse opinions and suggested principles from 27 individuals into consistent and readable form requires a significant amount of compilation. Unfortunately, the compiling and editing process may tend to eliminate some of the information and meaning contained in the original contributions. The voting procedure for opinions and suggested general principles is also an imperfect, problematic process. Participants found that they had to reject or disagree with an opinion if they disagree or objected to any part of it. In other situations, they probably compromised because of another participant's strong reaction to an aspect of the scenario or to a suggested general principle. In fact, some of the principles were favored by only one participant. In any case, consensus was not the goal of this project. Its purpose was to identify important ethical issues and to have a group of thoughtful, responsible people in computer science, technology, and business discuss them.

This book is not a treatise on ethics; rather, it is the collection of opinions and results of discussions of a group of 27 individuals. The SRI study leader and staff influenced the outcome only to the degree of setting the limits and framework for the issues and methodology and in suggesting the original set of scenarios. They did not participate either in the voting or in expressing opinions in the subgroup sessions, but they did participate in workshop plenary discussions.

The project has raised many more questions than it has answered. Nonetheless, at this stage of developing applications of ethical principles, it is appropriate to be posing provocative questions and suggesting a range of answers. Over time, better answers will emerge.

Codes of Ethics and Other Uses of the Studies

Based on the 1977 study, several works besides the book, Ethical Conflicts in Computer Science and Technology, were produced to further the cause of good ethical practices. The ACM and the ICCP incorporated principles derived from the workshop into ethical codes. In addition, a self-assessment procedure dealing with ethics in computing was widely distributed by ACM as Self-Assessment Procedure IX. A workbook for students associated with the 1981 book was also produced and is still being used in many universities and colleges. A similar workbook is planned to accompany this book.

It is encouraging that ethical philosophers are directing their attention to the computer field. Two notable examples are Professor Deborah G. Johnson at Rensselaer Polytechnic Institute (Computer Ethics, Prentice-Hall, 1985) and Professor Carol Gould, Stevens Institute of Technology (The Information Web: Ethical and Social Implications of Computer Networking, Westview Press, in progress, 1988).

We hope that similar uses will be made of the materials contained herein and that this study will promote the sound application of ethical practices in computer science, technology, and business. We suggest that you, the reader, pause after reading each scenario to deliberate and reach your own conclusions. Then as you continue reading, you can compare your conclusions with those of the project participants. You may also wish to develop your own variations of some of the scenarios.

Acknowledgments

My two coauthors, Dr. Bruce Baker and Ms. Susan Swope, have contributed greatly to the success of this book and the study on which it is based. Bruce was instrumental in the administration of the study, selecting participants, coordinating the workshop, and contributing his insights on ethics from his many years of university teaching. Susan was the compiler, organizer, and analyst who really put the book together, and she is the primary author of Chapter VI, "Summary of Ethical Issues."

Drs. Deborah G. Johnson and Susan Leigh Star were the acknowledged experts on the study team in ethical philosophy and sociology. They played the important roles of advisors, assisting throughout the study with ideas and prior work as they critiqued our scenarios, methodology, and the draft material for this book.

The 35 invited participants, especially the 27 workshop attendees, were responsible for the study being practical and representative of the thoughts about ethics in the computer field. In a sense, they are the real authors of this book. The opinions, positions, and principles reported are theirs. They also revised scenarios, created new ones, and reviewed the draft of the scenario chapters (II through VI). They are to be commended for enduring the discipline forced on them to address so may ethical issues in so short a time.

Barbara Stevens provided the editorial help to coalesce many different writing styles into a coherent whole.

Finally, our thanks to Dr. Rachelle Hollander, the National Science Foundation program officer for this grant, who contributed her expert guidance on grant matters, was enthusiastic about pursuing this new study, and added as much to our deliberations and expression of results as any other participant.

Donn B. Parker
SRI International

II
Professional Standards, Obligations, and Accountability

As an occupation evolves into a profession, the associated knowledge and skills expand, the requirements for entering the profession increase, and the standards of conduct required of members in good standing become more demanding. Although the computer science and data processing fields have not yet achieved the status of the medical, law, and accounting professions, computer and data processing standards are moving closer to those in the well-established professions. Such standards of conduct, obligation, and accountability should guide both neophytes and seasoned practitioners in the field.

People sometimes take questionable routes to achieve their or their employer's objectives, however. Students challenged to compromise their school's computer system, for example, may not question the acceptability of such activities. Employees may be asked to take shortcuts or ignore deficiencies that could potentially cause financial, emotional, or physical harm to the company, other employees, clients or customers, or the general public. In other cases, individuals may decide to market products or services that they know may not perform as promised.

In considering scenarios that explored the concepts of professional standards, obligations, and accountability, attendees raised many important questions about who is responsible and who has authority. For example:

- To what extent is a professional person responsible for reporting another's abuse of a computer system once made aware of it?

- Having learned of a vulnerability in a system, and knowing that others are aware of it, is one obliged to correct that vulnerability?

- Who can or should be held responsible for damages or injuries caused by a miscalculation, prank, or programming error?

- Are a tool's developers responsible for how it is used? Should they be held responsible for the tool's proper functioning?

- To what degree is it unethical to deceive or mislead others about the characteristics or functioning of a product?

- Under what circumstances should a manager investigate reports of unusual or questionable activities?

People relate to computers in fundamental roles — for example, as user, creator, seller — and as part of other roles — as student, professor, employee, entrepreneur. In these roles, people have goals, responsibility, and authority. Their overall responsibilities and accountability imply certain courses of action (and prohibit others) in their use of computers.

Such responsibility and authority carry with them the burden of accountability. Although professionals cannot be held accountable for areas in which they have neither expertise nor control, by accepting an assignment, whether within a company or for a client, a professional also accepts responsibility for the results of the performance of that assignment.

Only professionals, however, are expected to adhere strictly to standards of professional ethics when these standards require more than the standards of general morality that bind everyone. Students should observe the code of ethical conduct of the profession for which they are preparing. (Should they falter, however, they could perhaps be judged more leniently than they would be after graduation.) Professional educators (whether professors or directors of university computer service facilities) should encourage the observance of ethical conduct. Fostering cynical, immoral, or amoral attitudes among computer users is dangerous. Positive instruction and personal example should be used to teach that computers are valuable tools to be used properly, not abused.

Professionals are obligated to perform their work to the best of their ability and to insist on having the tools needed to carry out a task: Knowingly and intentionally producing inferior work is unethical, as is taking on work for which one is unqualified. Students may be tempted to do this because they often overestimate their knowledge and abilities. Although taking on work that brings new challenges is acceptable (for that is how we grow), it should be under circumstances that provide sufficient supervision to avoid any potentially harmful consequences.

The cases that follow explore some of the areas in which professionals confront ethical conflicts in their everyday work.

SCENARIO II.1[1]

STUDENT, SERVICE DIRECTOR: EXPLOITING VULNERABILITIES IN A UNIVERSITY COMPUTER SERVICE

A university student used the campus computer time-sharing service as an authorized user. The service director announced that students would receive public recognition if they successfully compromised the computer system from their terminals. Students were urged to report the weaknesses they found. This created an atmosphere of casual game playing and one-upsmanship in attacking the system.

---

[1] Scenario I.1 in the 1981 book, *Ethical Conflicts in Computer Science and Technology*.

The student found a means of compromising the system and reported it to the director. However, nothing was done to correct the vulnerability, and the student continued to use her advantage to obtain more computer time than she was otherwise allowed. She used this time to play games and continue her attacks to find more vulnerabilities.

Student: Using Computer Services by Taking Advantage of a Vulnerability

	Total	Unethical	Not Unethical	No Ethics Issue
1987				
General Vote	25	20	1	4
Subgroup Vote	6	4	0	2
1977				
General Vote	28	20	6	2
Subgroup Vote	12	10	1	1

Opinions

The 1977 group believed the student's action was dishonest, notwithstanding her motivation and the expectations of the service director. Even though the director failed to correct the vulnerability, the student should have requested additional computer time through regular channels. She was unethical when she continued the attacks and used computer time for game playing.

Some of the 1977 subgroup members believed the student was justified in taking advantage of the additional time to look for other vulnerabilities because of the director's explicit announcement that whatever could be wrong with the system should be uncovered and reported. Thus, in using this time to expose further vulnerabilities, the student was still acting to achieve an accepted purpose.

In 1987, group members' opinions of the student's actions ranged from outright theft to mildly unethical. One participant, however, indicated that he believed the student was simply continuing to play the game the director started and thus was not acting unethically. Another suggested that the student's behavior cannot be judged within the norms of professional ethics, those being more appropriate for adults or professionals, and that general concepts of right and wrong are more appropriate.

Service Director: Encouraging Compromise of the Computer System

	Total	Unethical	Not Unethical	No Ethics Issue
1987				
General Vote	25	9	7	9
Subgroup Vote	6	2	1	3
1977				
General Vote	17	9	3	5
Subgroup Vote	10	2	3	5

Service Director: Not Correcting the Vulnerability

	Total	Unethical	Not Unethical	No Ethics Issue
1987				
General Vote	25	18	3	4
Subgroup Vote	6	5	0	1
1977				
General Vote	15	8	3	4
Subgroup Vote	11	9	1	1

This issue raises the question of whether students should be expected to comply with professional ethics (since their educational goal is toward that end) or whether a more forgiving student ethic should be applied that ends at graduation. Yet, what significance does the graduation event have relative to ethical behavior, especially considering the numbers of adult students and students engaged in professional work before graduation? Is the learning experience weakened by leniency? These are important issues that are certainly not new or unique to education.

Opinions

In both 1977 and 1987, participants strongly agreed that the service director created a "beat the system" challenge and atmosphere by not correcting the weakness in the system. As one said, "He left the keys in the car for the students." Although the service director had not committed himself to correcting deficiencies discovered by the students, his behavior was nonetheless unethical. His responsibility was to maintain a system that would not be jeopardized by misuse and to make it as difficult as possible for a user to cheat the system. The director's failure to act on the information the student gave him was a failure to fulfill his responsibility. His failure to correct vulnerabilities promoted student cynicism and abuses of the system. He should have used other, less dramatic ways of finding vulnerabilities. Another participant indicated, however, that the director was not required or expected to address all known vulnerabilities.

Changes in the computer world from 1977 to 1987 affected participants' responses. For example, the rise and visibility of the hacker culture and of computer intrusions have made computer service directors more sensitive to system vulnerabilities. On the other hand, the ready availability of computer services today makes them less precious than they were in 1977, and the attitude about accounting for usage is more casual. Therefore, although the voting profiles are very similar, the rationale underlying the votes may have changed between 1977 and 1987.

General Principles

1977 — The existence of temptation does not justify irresponsible action. The absence of locks does not justify theft for example.

Reporting knowledge of a system weakness does not remove the ethical responsibility to refrain from exploiting it.

Computer service resources should be conceived and used with care, as are those of a library, laboratory, or other important service facility.

Managers of computer services have a professional obligation to actively discourage users from treating the computer as a toy and using it for self-promotion. By doing so, they reduce amorality and cynicism in computer use.

Encouraging users to attempt to compromise a system leads to situations inviting unethical behavior.

1987 — Verbal or written contracts, whether rules for games or exchange of goods and services, can greatly reduce the opportunity and motivation for unethical behavior.

Teaching ethical values by encouraging negative experiences is a poor and dangerous technique.

SCENARIO II.2[2]

ENGINEERS, SYSTEMS, ANALYSTS, AND PROGRAMMERS IN A STATE AGENCY: ASSUMING LEGAL RESPONSIBILITY FOR WORK

Civil engineers employed by a state agency were engaged in numerous construction design projects, such as flood control, where safety of humans is a factor. They were held personally responsible for their work, under a professional and business

---

[2]Scenario I.7 in the 1981 book.

responsibility law. In their design activities, the engineers increasingly relied on computer programs that were designed by systems analysts and implemented by computer programmers. The engineers specified the problems requiring solution and, to various degrees, specified the methods of solution and test cases for demonstrating that the computer programs functioned correctly. Several of the computer programs included logic where decisions were based on engineer-specified criteria and where the program output selected types and quantities of construction materials and stated how deliverable end products were to be constructed.

The engineers complained to their management that they were not able to determine the correctness and integrity of the computer programs, and the results of their work relied heavily on those qualities. Therefore, an error in a computer program or an error in operation of the computer (that could be detected by the programmer) could result in a serious design flaw that could cause harm to people. The engineers wanted the systems analysts and computer programmers to share the responsibilities for any losses under the professional and business law.

The systems analysts and programmers stated that they were merely providing tools and had no involvement in their use. The engineers could test and analyze the programs to assure themselves of their accuracy. Therefore, the systems analysts and programmers should not be held responsible.

System Analysts and Programmers: Resisting Acceptance of Responsibility

	Total	Unethical	Not Unethical	No Ethics Issue
1987				
General Vote	26	5	18	3
Subgroup Vote	7	1	6	0
1977				
General Vote	22	9	7	6
Subgroup Vote	7	5	0	2

Opinions

In 1977, a plurality of participants agreed that the systems analysts and programmers should work more closely and share responsibility with the engineers to ensure that the programs adequately addressed public safety needs. Yet they acknowledged that the computer personnel could not assume the engineers' role. Their expertise is in taking information supplied to them and constructing it into a cohesive whole that aids and simplifies manipulation of the information. As long as the computer personnel followed the agreed on program specifications, they did their job adequately and could not be held responsible for any engineering problems that might result; without engineering expertise, they could not be expected to decide whether the outputs of their programs were consistent with engineering standards of safety. Thus, expecting them to be responsible for the engineering aspects of a project, such as design or material selection, would be unreasonable. They are responsible only for ensuring that their programs meet the requirements of good computing.

While acknowledging these arguments, the 1987 group went further. Generally speaking, systems analysts and programmers are not licensed, certified professionals in the same sense as civil engineers. Because the professionally certified engineers supplied all the data, including test data to check the programs, they were responsible for using professional judgment and care to be sure they provided accurate data and thoroughly checked the results of others' work. Using the engineers' rationale, many others besides the analysts and programmers might be held responsible for errors, including computer operators.

Some of the 1987 participants argued that the computer programmers and analysts took a weak and self-serving position. They suggested that the engineers would have to become competent computer analysts and programmers to be able to ascertain the correctness of the programs. However, if the engineers could test and analyze the programs, they would not require the services of computer personnel. They contended that the computer personnel were unethical because they were providing tools but trying to detach themselves from the use of those tools. They should be held responsible for the integrity of their programs. The computer personnel were the best judges of the adequacy of the tools and therefore could not avoid responsibility for errors arising from their application. The argument that they were merely providing tools and had no involvement in their use is untenable. The concept of tool logically implies use — in this case, the construction of safe engineering projects.

Attendees maintained that if the professional and business responsibility law is sufficiently broad, neither group should be able to escape responsibility for its acts. Certainly a mistake in a program can be evaluated as easily as a mistake in bridge construction. Both groups should be judged by the standard of exercising reasonable judgment in light of all circumstances. Computer personnel have at least a limited ethical liability where public safety is concerned, because they are involved with technical matters that go beyond the engineers' function and expertise.

The 1987 subgroup believed it is not unethical for the computer personnel or the engineers to argue their particular positions. Ethical issues arise when commitments are, or fail to be, made. A violation of contracted accountability would be unethical. As a matter of business practice in a dispute, refusal to accept accountability before any inquiry occurred would not be unethical. Others said no ethical issue existed in that a person has a right to attempt to refuse anything before making any commitment.

Engineers: Contending Analysts and Programmers Should Share in Legal Responsibility				
	Total	Unethical	Not Unethical	No Ethics Issue
1987				
General Vote	26	7	10	9
Subgroup Vote	7	0	7	0
1977				
General Vote	17	7	8	2
Subgroup Vote	6	2	3	1

Opinions

Most participants in both 1977 and 1987 agreed that the engineers should be held responsible for their work, regardless of whether computers are used. Even though they did not possess the expertise necessary to judge the adequacy of a computer program, they should know whether the outputs conform to design and construction standards. The engineers would have to make these judgments by some other means if they did not have access to computer programs to simplify the task. To demand that others share responsibility for a task where engineers alone possess the expertise to judge is consequently neither reasonable nor morally acceptable. The engineers should have teamed to test the validity of the programs adequately and not try to avoid their legal responsibility. Technological ignorance is no excuse for malpractice.

A majority also agree that the engineers' desire to share a portion of the legal responsibility was not unethical. They were responsible for providing the computer personnel with correct problem statements and test cases. They could not, however, prove the validity of the programs any more than they could test every rivet in a bridge built by the contractor. The engineers were responsible for the design, and others were likewise responsible for their roles.

As some participants noted, whether engineers could determine the correctness and integrity of the computer programs relates to their technical background. A related question is whether engineers on a given project have the time or resources to test and analyze computer programs — assuming they could understand them. If the engineers cannot understand the programs, systems analysts and programmers should share a legal responsibility as an incentive to accuracy and correctness.

One can delegate authority, but not responsibility. The engineers were responsible for the final construction; even if the computer personnel had been made responsible under the professional and business law, their responsibility would have been limited to following the engineers' specifications for methods of solution and tests. In those cases where the engineers did not give complete specifications, they would be derelict in their duty and therefore unethical. One 1987 participant added that the engineers were unethical because they were trying to rid themselves of an already accepted accountability, but they might have some legal recourse against the computer people in the case of a specific dispute.

The 1987 participants identified important issues at stake, including public safety, professional responsibility in the broadest sense, guarantee of one's work for others, and the great differences between disclaimers before and after the fact from a legal standpoint. Increasing numbers of disputes about shared responsibility and their settlement are signs of the maturing of the computer profession relative to other long-standing professions.

General Principles

Analysts and programmers should be responsible for a program's performance according to specifications. Engineers should be responsible for the specifications.

A program is a tool, but that does not mean that the tool's designer has no responsibility for its use. If a manufacturer contracts with a machine designer for a machine made to certain specifications, including certain safety features, the designer must be held responsible if the machine does not perform according to the claims made for it.

If computer analysts and programmers will not share responsibility for computer work, they should not be hired. The work of developing programs should be given to a company or individual with both computer and engineering expertise so the two parties can jointly accept responsibility. The assignment of responsibility is one of the most difficult problems in law or ethics.

Where significant technical expertise is needed, those with that expertise, not the end user, must be responsible for their work.

Systems analysts and programmers must voluntarily accept (or be required to accept) responsibility, and they must be given authority in their sphere of work. If they have the responsibility and the authority, they are also forced to accept accountability.

Systems analysts and programmers should share with the users of their product a portion of the legal responsibilities that bind the users. How the legal responsibilities should be divided is a legal question.

SCENARIO II.3[3]

PROGRAMMING MANAGER, COMPANY: IGNORING INDICATIONS OF FRAUD

The manager of a computer applications programming department in a large company is responsible for all applications programming throughout the company. Most of the programmers report to him. Several people in other departments also program, but must conform to the standards in the programming department. All computer application programs that are to be run on a continuing production basis must be tested and approved by the quality assurance group in the manager's department before they go into regular production use.

By inflating company assets, the top executive officers of the company are engaged in a massive fraud against the stockholders and other investors. Significant evidence of the fraud is contained in the data files stored and processed by the computer. Several computer programs have been developed to assist in the fraud. A programmer in the inventory control department and a programmer in the applications programming department are knowing participants in the fraud. The programs used in the fraud perform unorthodox functions that obviously do not contribute to legitimate business functions. For example, one program causes statistically selected inventoried items, which are recorded in

---

[3]Scenario II.3 in the 1981 book.

master files, to be declared obsolete and ready for disposal. Another program generates inventory records of nonexistent items by copying and slightly modifying records of existing items. These programs have passed quality assurance testing on the pretext that they would be used to generate test master files for use in business simulations for the planning department. Released to the computer operations department for production, they were actually used to modify the master files of the company.

The programming manager was unaware of the involvement of either programmer. The fraud-related programs were not suspicious to him. Several programmers reported to him that unexplained, unusual activities were going on in several parts of the company. He was aware of the company's severe financial problems and was told to do only what he was asked to do without question, and not to bother about the implications, because the problems would soon be solved.

The fraud was ultimately discovered and the perpetrators prosecuted. The programming manager was identified as an unindicted coconspirator.

Programming Manager: Not Responding to Evidence of Wrongdoing

	Total	Unethical	Not Unethical	No Ethics Issue
1987				
General Vote	24	23	1	0
Subgroup Vote	7	7	0	0
1977				
General Vote	25	10	10	5
Subgroup Vote	11	7	4	0

Opinions

Although 1977 group members did not concur about the ethicality of this issue, by 1987 participants nearly unanimously agreed that not responding to evidence of wrongdoing was unethical. No one could identify any changes since 1977 that would evoke such a different perspective, however.

In 1977, participants could not reach a consensus, possibly because the scenario did not clearly describe the manager as being informed of the fraud. Group members agreed that if the manager could have learned of the fraud, or actually knew of it, he was unethical for not acting. The phrase, "unexplained, unusual activities," is vague and does not connote any urgency or impropriety. Nevertheless, he had a responsibility to inform the auditors of possible problems since they had a wider range of responsibility. The manager also could have set ethical standards for all programmers. (Nothing in the scenario says he did not.) Although participants believed he had not acted unethically, neither had he taken any positive steps to maintain an ethical posture.

Several of the 1977 group also noted the difference between ethics and legal responsibility. The manager in this case did nothing unethical, but he was accountable and thus correctly identified as an unindicted coconspirator. Others disagreed, contending that he did not conspire with anyone; he did not join in a secret agreement to do an unlawful or wrongful act. To conspire is to act; he failed to act. He therefore should not have been identified as an unindicted coconspirator because he did not have direct knowledge and did not stand to benefit. No ethics issue was involved; the manager was told only of some unexplained, unusual activities and trusted his boss.

Several of the subgroup asserted that not he, but his supervisor, acted unethically, on the principle that a supervisor is responsible for the actions of those under him and the programming manager was not a line manager over the programmers committing the fraud. However, the subgroup was equally split over the plausibility of the manager's explanation for the unusual conduct. Those accepting this explanation maintained that he was not suspicious of the programmers; therefore, he could not be blamed for trusting his boss and his behavior was not unethical.

Willful ignorance, softened somewhat by a manager's likely lack of knowledge of application program details, is a more plausible explanation than unintended ignorance. If the manager could not assess the programs in question, he could have instructed his quality assurance group to examine them and their use. Apparently, he preferred not to know too much, for he would otherwise be under a moral obligation to blow the whistle or resign. His action appears to be self-protective. This attitude kept him out of jail, but it does not morally exonerate him.

Others contended that the manager would have to be stupid not to see connections among the events described. Stupidity, however, is not a crime, and no generally accepted standards for professional competency among computer programmers cover this case.

A differing minority agreed that the manager was unethical because of culpable ignorance. He was responsible for all use of the computer services. He failed to investigate the complaints of his subordinates and suspicious circumstances; he was aware of financial problems; and if he had to be told to mind his own business he apparently was suspicious.

On the other hand, the programs had a legitimate use. The fraud involved their misuse, for which the programming manager was not responsible. Neither was he responsible for investigating reports of unexplained, unusual activities in other parts of the company. His employer made this clear. Perhaps he should have alerted the managers who were responsible for the areas in which the unusual activities took place. Determining where one's ethical responsibility for the acts of others begins and ends is sometimes difficult.

The majority disagreed with the opinion that the manager could not be faulted for carrying out company directives, assuming he was competent and that his judgment would have been supported by his peers. The arguments introduced in 1987 were essentially the same as those offered by the 1977 group, with virtually all the 1987 group concurring that the programming manager acted unethically.

General Principles

Computer personnel have the ethical responsibility to inform appropriate authorities when they are suspicious of computer misuse.

Incompetence is not unethical. An ethics issue arises when managers fail to resolve doubts based on their knowledge of suspicious activities.

The number of programmers and computer programs in a manager's organization is no criterion for determining or limiting his responsibility for them.

Persons in responsible positions are obligated to know what is happening in their departments or divisions. Ignorance or even avoiding knowledge does mitigate culpability somewhat, but doing only what one is told is never an acceptable excuse.

An act of omission is unethical only when wrongdoing or harm is suspected.

SCENARIO II.4[4]

PROJECT LEADER, MANAGEMENT, RETAIL COMPANY: INSTALLING AN INADEQUATE SYSTEM

A programming analyst was given project responsibility to develop a customer billing and credit system for his employer, a

---

[4]Scenario IV.4 in the 1981 book.

large retail business. He thought the budget and resources he was given were adequate. However, the budgeted amount was expended before completion of the system. He had continually warned management of impending problems, but was directed to finish the development as soon as possible and at lowest cost. He was forced by management to do this foregoing many of the program functions, including audit controls, safeguards, flexibility, error detection and correction capabilities, automatic exception handling, and exception reporting. A "bare bones" system was installed. He was told that he could add all the omitted capabilities in subsequent versions, after production of the initial system.

A difficult, expensive, and extensive conversion to the new system occurred. After the new system was in production, great problems arose. Many customers received incorrect and incomprehensible billings and credit statements and became outraged. The retail company was unable to correct errors or explain confusing system output. Fraud increased. Business and profits declined, and customers suffered much anguish and personal expense. The project leader was blamed for the losses.

Project Leader: Implementing an Incomplete and Inadequate System				
	Total	Unethical	Not Unethical	No Ethics Issue
1987 General Vote	26	18	4	4
Subgroup Vote	9	5	3	1
1977 General Vote	31	9	14	8
Subgroup Vote	7	6	0	1

Opinions

Between 1977 and 1987, many more participants concurred that the project leader acted unethically. Management did force him to produce an inadequate system, although the project leader continuously warned management of the problems this act would cause. Nevertheless, because professionals are obligated to do the best work they can do, the project leader in this situation should have refused to do the work and even resigned if necessary.

The minority accepted the budget constraints placed by management and considered the project leader's warning to have absolved him of responsibility for the consequences.

Management: Ordering the System Into Production Prematurely				
	Total	Unethical	Not Unethical	No Ethics Issue
1987				
General Vote	27	24	0	3
Subgroup Vote	7	7	0	0
1977				
General Vote	12	11	1	0
Subgroup Vote	8	7	0	1

Opinions

Group members' opinions remained essentially the same between 1977 and 1987, namely, that management acted unethically. By not correcting known problems and implementing an inadequate system, they violated their fiduciary duty both to their customers and to their stockholders or investors. Three people in the 1987 group disagreed, however, saying that the order involved mismanagement, or a lack of judgment, but not ethics.

Management: Blaming the Project Leader			
	Total	Unethical	Not Unethical	No Ethics Issue
1987				
General Vote	26	24	0	2
Subgroup Vote	7	7	0	0

Opinions

The 1987 group separated blaming the project leader from prematurely ordering the system into production. They believed management acted unethically in blaming the project leader. Those who voted that no ethics issue was involved argued that the project leader had not used good judgment and that the reprimand was a direct result of his shortcomings.

General Principles

Professionals have a responsibility to do good work and to be accountable for that work. They should attempt to remove any impediments to properly doing the job, including bringing problems to the attention of a higher authority, protesting unreasonable constraints, whistle blowing to an outside agency, or even resigning.

The lengths that professionals must go to maintain their standard of ethics depend on the degree of certainty of the inadequacies and the severity of the damage that will be caused if those inadequacies are not corrected.

SCENARIO II.5
COMPUTER SCIENTIST: ACCEPTING A GRANT ON A POSSIBLY UNACHIEVABLE PROGRAM

A professor of computer science applied for and received a grant from the Strategic Defense Initiative Program to engage in a software assurance research project of a theoretical nature. The goal was to determine the methods by which error-free software might be produced on a large-scale basis. The professor does not believe that SDI is a viable Department of Defense program. She does believe, however, that her work could add measurably to the body of scientific knowledge concerning the development of error-free software. Thus, she accepted the grant money.

Professor: Accepting a Grant for Reasons She Found Compatible with Her Value Judgments				
	Total	Unethical	Not Unethical	No Ethics Issue
General Vote	23	14	7	2
Subgroup Vote	6	3	3	0

Opinions

Half of the subgroup and most of the general group considered it unethical for the professor to accept grant money for a segment (development of error-free software methodology) of a project (SDI) that she believed would fail. Their reasons for considering the acceptance unethical are that:

- Acceptance indicates at least an implicit endorsement of a project she believes will not work by a professional who is likely to influence others.

- This acceptance sends what the professor believes to be an erroneous message (SDI is viable) to the general public.

- She is being dishonest with the funding agency, which is unaware of her opinion.

These participants concluded that the professor should find some other way to fund her research.

The minority contended that her action was not unethical because:

- She questioned only the viability, not the morality, of SDI.

- She did believe that her project was both viable and valuable for scientific knowledge apart from its use in the SDI program.

- She does not know whether SDI will work or not. (Neither does anyone else.)

- She believes she is acting ethically.

Therefore, they argue, belief in SDI itself is irrelevant as long as the professor believed that she could deliver the product for which she accepted the grant. On balance, even with the implied SDI endorsement, one participant said the benefits of accepting the grant outweighed the disadvantages.

General Principles

Professionals are accountable to the general public and thus have a responsibility to avoid misleading them.

Professionals should not accept funding for research under false pretenses.

Professionals must be concerned with the ethics of the actions in which they are directly involved. They cannot be held responsible for the unanticipated possible misuses of their work by others over whom they have no control. They are obligated, however, to make their doubts and the limitations of their work known.

SCENARIO II.6

SOFTWARE DEVELOPER: RELYING ON QUESTIONABLE INPUTS

A software professional was assigned the task of developing software to control a particular unit of a large system. Preliminary analysis indicated that the work was well within the state of the art, and no difficulties were anticipated with the immediate task.

To function properly, or to function at all, however, the software to be developed required inputs from other units in the system. Someone gave the professional an article by an eminent software specialist that convinced him the inputs from other units could not be trusted. Thus, neither the software he was designing nor the unit his company was providing could correctly accomplish their task. The professional showed the article to his supervisor and explained its significance. The supervisor's response was, "That's not our problem; let's just be sure that our part of the system functions properly." The software professional continued to work on the project.

Software Professional: Working on a Project That Depends on Questionable Inputs				
	Total	Unethical	Not Unethical	No Ethics Issue
General Vote	20	12	7	1
Subgroup Vote	6	4	2	0

Opinions

The majority of participants maintained the software professional had a responsibility to go beyond his supervisor, if necessary, to call attention to the questionable inputs. Some discussion centered on the consequences of using these data. If the errors could cause serious harm, the professional is ethically required to do whatever is necessary to ensure that they are corrected. These actions would include notifying someone with broader system responsibilities, informing management or the public, and even resigning. To do otherwise in such a situation is unethical. If the consequences would be relatively trivial, the group agreed that telling the supervisor of his concerns might be sufficient

The minority opinion was that the software professional acted properly in notifying his supervisor of his concerns. He was not ethically required to do more. Some participants discussed the ramifications of whistle blowing; frequently, whistle blowers not only lose their jobs, but also are marked for life. System deficiencies that cause severe damage warrant taking such a risk, however.

Supervisor: Being Concerned with Only One Part of the System				
	Total	Unethical	Not Unethical	No Ethics Issue
General Vote	22	21	1	0
Subgroup Vote	6	6	0	0

Opinions

Nearly everyone believed the supervisor acted unethically if he did not inform management of the deficiencies once he became aware of them. It is unethical to produce a product known not to work properly. Both personal integrity and company loyalty demand that others be alerted to such system deficiencies.

The one dissenter contended that every person is not responsible for everything that happens. An individual can be held responsible only for what is under his control.

General Principles

Computer professionals must consider their certainty that a problem exists and the severity of the damage that may be caused. If certainty and severity are both high, professionals must go as far as necessary to see that the problem is addressed.

Professionals are responsible for the context, as well as the particulars, of the work they do.

Professionals should not agree to perform work that they know cannot be successfully completed.

SCENARIO II.7

OFFICER, PROFESSIONAL SOCIETY: NOT PUBLICIZING A CRIMINAL CONVICTION

A computer security consultant had been convicted of a felony computer-related crime and served a prison sentence. Since that time, he had held two responsible information systems management positions with highly regarded institutions that were well aware of his criminal past. For 5 years, the manager refused to talk to the news media or anyone else publicly about his crime or conviction. He became a member of a national professional society and was active in one of its major chapters; the national officers of the professional society knew of his criminal past. The manager ran for the office of chapter chairman and won the election. During the campaign, he had consciously not mentioned his criminal conviction in his election biography, consistent with his ongoing practice. After the election, another computer security consultant publicly and widely disclosed the new chapter chairman's background and complained to the national officers of the professional society. The issue received national public exposure, which embarrassed the professional society and the convicted individual, prompting public debate about the propriety of all the actions, positions, and events of the incident.

Security Consultant: Running for Professional Society Office Without Disclosing Felony Conviction

	Total	Unethical	Not Unethical	No Ethics Issue
General Vote	24	10	13	1
Subgroup Vote	6	2	4	0

Opinions

This case prompted substantial discussion on both sides. A slight majority of both the general group and subgroup decided that the security consultant was not unethical to hide his past. Although it would be unethical to deceive society members, no one is required to reveal all his past, in particular, negative information. The consultant had served his time, maintained a clean record since then, and told those who should know (employers, society officers) about his conviction.

A significant minority of the participants insisted that the information about the criminal record was germane to the position the consultant was seeking and those who voted in the election had a right to know. He should have fully disclosed his past. To do otherwise is dishonest and unethical. Public figures do not have the same rights to privacy as private individuals.

Professional Society: Not Removing Chairman After Criminal Past Becomes Known

	Total	Unethical	Not Unethical	No Ethics Issue
General Vote	24	7	12	5
Subgroup Vote	6	1	4	1

Opinions

Not removing the chairman from office prompted strong discussion. The majority agreed that the society did not act unethically. The national officers could reason that the consultant had paid his debt to society, reformed, and so could decide whether to reveal that particular part of his past during the campaign. There would be no justification for removing him unless the bylaws of the society prohibited anyone with a felony conviction from holding office.

Those who considered the society unethical in not removing the consultant from office said that the society had an obligation to its members. The officers should have insisted that the candidate fully disclose his past to the electorate. Several also questioned whether the society's bylaws might disbar convicted felons from holding office. One member added that the society should have called a new election.

Second Consultant: Publicizing the Criminal Record of the New Chairman After His Election

	Total	Unethical	Not Unethical	No Ethics Issue
General Vote	26	10	15	1
Subgroup Vote	7	5	2	0

Opinions

The general group and subgroup majorities disagreed about the ethicality of publicizing the felony conviction after the election. Most of the general group contended that such disclosure was not unethical. The consultant simply placed the information, a matter of public record, out in the open. One group member, in fact, stated that it was the second consultant's duty to reveal a computer-based crime that was relevant to the society. One who voted not unethical still questioned the consultant's motives in waiting until after the election.

The substantial minority who believed the disclosure was unethical reasoned that release of the information was unjustified, especially after the election. One participant argued the only likely motive for publicity was malice. The reputations of both the consultant and the professional society were damaged. Another called the action unprofessional, believing that a hearing among the society leadership would have been more proper.

General Principles

Public figures cannot expect to maintain the same degree of privacy as private citizens.

Full disclosure is the best policy.

Muckraking and sensationalism are never justified.

SCENARIO II.8

OWNER, COMPUTER DATING SERVICE: NOT TAKING RESPONSIBILITY FOR RISK TO CUSTOMERS' HEALTH

An entrepreneur went into the dating service business, using a computer to match couples. Several customers contracted AIDS and suspected that they acquired the disease from a person they had met through the dating service. They demanded that the entrepreneur terminate his business. He refused, indicating that social diseases are always a risk to sexually active people and that he has no responsibility to warn them or to determine health risk factors in accepting and matching customers.

Customers with AIDS: Demanding That Computer Dating Service Owner Terminate Business

	Total	Unethical	Not Unethical	No Ethics Issue
General Vote	24	3	6	15
Subgroup Vote	7	0	0	7

Opinions

The entire subgroup and most of the general group believed the demand to stop operating the service, per se, was not an ethical issue. In demanding the service be terminated, the customers were within their rights under the first amendment. The only ethical question is whether the owner must accede to the demand. Those who voted that the demand was not unethical presented essentially the same argument.

A few participants contended that such a demand was unethical. One questioned the strength of the customers' suspicions and whether they had any evidence to support those suspicions.

Computer Dating Service Owner: Refusing to Warn Clients or Determine Health Risk Factors

	Total	Unethical	Not Unethical	No Ethics Issue
General Vote	26	3	20	3
Subgroup Vote	7	0	7	0

Opinions

Most group members believed the entrepreneur was not acting unethically in refusing to take action. He was correct in stating that the law does not require him to warn clients about possible health risks. Individuals freely choose to use a dating service, and dating alone does not cause AIDS. Most people today know the risks involved in being sexually active. Although some participants thought that the business owner should issue a disclaimer, he is not obliged to do so.

A few asserted that being in such a business today is unethical (because of the prevalence of diseases such as AIDS). By continuing the services, the owner has a moral responsibility to warn people of possible life-threatening risks and to refuse to accept customers known to have AIDS.

General Principles

It is not unethical to offer a service that has potential risks as well as benefits, as long as the risks are known to those who choose to use that service.

Where risks are life-threatening, those offering services may have a moral responsibility to remind customers of the need to take precautions.

SCENARIO II.9

GOVERNMENT AGENTS, ENTREPRENEUR, STUDENT EXPERT: EXECUTING A WARRANT TO SEARCH PROPRIETARY, COMPUTER-STORED DATA

An entrepreneur established a small business to produce financial statements for people applying to banks for loans. He stored all his business records and reports for his clients on a 20-megabyte hard disk in a microcomputer system.

Federal agents investigating suspected criminal activities by some customers of the entrepreneur obtained a search warrant to obtain the financial records of these people. Unfamiliar with microcomputers, the agents hired a student in the computer science department of a local university who claimed he was familiar with the type of microcomputer being used. The agents staged a surprise raid on the entrepreneur's business. They instructed the student to examine the contents of the hard disk and print out the financial statements of the people named in the search warrant. In doing this, the student accidentally erased the entire contents of the hard disk.

Because the entrepreneur had no backup of the contents of the microcomputer, he lost all his business records and his business collapsed. The agents failed to obtain the information necessary for their investigation. The student returned to his studies and was not paid for his efforts.

Federal Agents: Failing to Obtain Appropriate Expertise and Thereby Causing a Business to Lose Records				
	Total	**Unethical**	**Not Unethical**	**No Ethics Issue**
General Vote	24	22	1	1
Subgroup Vote	6	6	0	0

Opinions

All subgroup members and nearly all the general group believed the federal agents to have acted unethically in not ensuring that the hired expert could search and print the entrepreneur's files without damaging them. The agents had a professional obligation to ensure that the job was performed properly. Moreover, police should be held accountable for destruction of a third party's property. At the very least, they exercised poor judgment in this case.

Two participants had differing opinions. One argued that the agents may have been unaware of the danger to the records. They may have thought that they had used reasonable care in acquiring expertise to access them, and therefore, no ethics issue was involved. In arguing that the agents did not act unethically, another participant said that the incident was simply an unfortunate accident. At the same time, he admitted that the agents should have chosen the expert more carefully.

Entrepreneur: Failing to Back Up Data				
	Total	Unethical	Not Unethical	No Ethics Issue
General Vote	25	3	0	22
Subgroup Vote	6	0	0	6

Opinions

Everyone in the subgroup and the great majority of the general group agreed that failing to back up data was not an ethics issue. It was stupid, perhaps negligent, of the entrepreneur not to back up his data, but not unethical. Anyone whose business depends on computer data should certainly know the importance of taking such precautions.

The few who considered the entrepreneur's action unethical considered that his failure might affect clients whose financial records were destroyed. He should have protected his clients' interests; to do otherwise is unethical.

General Principles

To damage property by neglecting to exercise due diligence is unethical.

Business people have a responsibility to protect their clients' interests.

Police have a responsibility not to harm the innocent or even those suspected, but not convicted, of misdeeds.

SCENARIO II.10

PROGRAMMING ANALYST: QUESTIONING THE ETHICS OF HONESTY TESTING

Recruiters in the human resources department of a large company administered an honesty test to job applicants using a booklet of questions and paper answer forms. Because they had so many applicants for so few jobs, they routinely eliminated all applicants whose responses did not fit statistically into the safest range.

A programming analyst was assigned the task of developing an on-line computer-based test that would allow for much greater real-time discretion by recruiters. The belief was that this would probably improve the effectiveness of the testing to the extent that more dishonest people and fewer qualified, honest people would be screened out.

The programming analyst refused to work on this project, claiming that any automatic screening for honesty was an intrusion into a person's privacy. The human resources manager argued that automating the screening would provide much fairer evaluations and results and would significantly reduce the disclosure of personal information to the recruiters, since they would examine only the exceptional cases directly. These arguments convinced the programming analyst that the new computer application was far more ethically sound than the old manual method, and he agreed to proceed with the job.

Human Resources Manager: Asking Programming Analyst to Develop a Computerized Honesty Test			
	Total	Unethical	Not Unethical	No Ethics Issue
Subgroup Vote	7	1	5	1

Opinions

Most subgroup participants agreed that computerizing the honesty test to give recruiters more real-time discretion was not unethical, based on the following assumptions:

• Applicants knew they were being tested for honesty

• The test itself was accurate — it worked, and it was free from human biases that would skew the results.

The general group, however, decided not to vote on this scenario. They were concerned that use of a computer creates an aura of objectivity that often is not present. The people who construct the test content, consciously or unconsciously, include certain biases that are sometimes reflected in the content. The test could, for example, be designed to screen prospective employees for pro- or anti-union proclivities.

One participant, familiar with many tests of this type, found them all extremely skewed and, in his opinion, invalid. Because he did not believe an accurate test could be constructed, he contended that administering honesty tests was immoral and unethical.

Another participant reasoned that individuals who were dishonest could and well might understand the evaluation process and lie on the test. As a result, they might pass the test easily and be thought honest whereas a basically honest person, admitting some transgressions, might fail the screen. Whom is the employer better off hiring? Moreover, the participant viewed the test as an invasion of privacy and argued that the use of this kind of test should be outlawed.

Still others were concerned that they were not sufficiently knowledgeable about these testing methodologies to judge the ethical permissibility of using them at all. They considered the scenario issues too important to vote with any confidence. The issues need to be explored further.

Programming Analyst: Agreeing to Proceed with Program Development			
	Total	**Unethical**	**Not Unethical**	**No Ethics Issue**
Subgroup Vote	7	1	6	0

Opinions

All but one subgroup member voted that the programmer was not acting unethically in proceeding with program development; he had questioned the practice and was convinced that the human resource manager's arguments were sound. If he had not been persuaded that the test would be much more fair, he should not have participated in its development.

General Principles

It is an invasion of privacy, and therefore unethical, to give individuals a test designed to reveal character without informing them of the test's purpose.

It is unethical to administer and use the results of a test that are known to be inaccurate.

SCENARIO II.11

SOFTWARE COMPANY: IGNORING VOTING MACHINE MALFUNCTIONS

Company XYZ has developed the software for a computerized voting machine. Company ABC, which manufactures the machine, has persuaded several cities and states to purchase it; on the strength of these orders, it is planning a major purchase from XYZ. XYZ software engineer Smith is visiting ABC one day and learns that problems in the construction of the machine mean that one in ten is likely to miscount soon after installation. Smith reports this to her superior, who informs her that that is ABC's problem. Smith does nothing further.

Software Engineer: Not Going Beyond Her Immediate Supervisor				
	Total	Unethical	Not Unethical	No Ethics Issue
General Vote	24	23	1	0
Subgroup Vote	7	6	1	0

Opinions

Group members nearly unanimously agreed that doing nothing further would be unethical. When there is certainty of defects, and when those defects could result in serious damage, professionals are obliged to take whatever action is necessary to see that the deficiencies are corrected. Use of inaccurate voting machines could invalidate elections and potentially harm the general public. Even on a pragmatic level, Company XYZ's best interests would not be served by having its software used in inaccurate machines. Its reputation and future sales may also be damaged when the deficiencies are uncovered. Responsible (ethical) behavior and good business practice are not inconsistent. The software engineer should pursue the matter further. The one individual who disagreed with this view believed that Smith's reporting the problem to her superior was sufficient, that no further action was necessary.

Superior: Telling Smith to Ignore the Malfunctions				
	Total	Unethical	Not Unethical	No Ethics Issue
General Vote	24	24	0	0
Subgroup Vote	7	7	0	0

Opinions

Everyone agreed that the superior acted unethically in failing to follow up on the reported deficiencies. Professionals have a duty to avoid, or at least try to avoid, causing harm to the public. One group member maintained that software developers are in a privileged position to detect, and thus prevent, certain kinds of computer malfunctions, even when they are not themselves the cause. They therefore have a special responsibility to their clients and to society to ensure that products perform the tasks for which they were designed.

General Principles

When a higher good is at stake in solving a problem, professionals must accept responsibility for prudently pursuing solutions.

Professionals have a duty to avoid, or at least try to avoid, causing harm to the public.

The extent of the professional's responsibility to solve a problem with a product depends on the degree of certainty that the problem exists, the cost of solving or avoiding the problem, and the severity of the potential damages if the problem is not corrected. When there is certainty of defects and when those defects could result in serious damage, professionals are obliged to take action to see that the deficiencies are corrected.

(See also position paper in Appendix.)

III

Property Ownership, Attribution, Piracy, Plagiarism, Copyrights, and Trade Secrets

Our society has long recognized that taking or using another's property without permission is wrong — whether that property is an idea or a physical object. Thus, the copyright was developed to protect written or intellectual property, even as deeds, patents, and other instruments were developed to protect physical property. With the advent of the computer, we have acknowledged the principle that software is a kind of intellectual property and that to copy or use it, without the owner's proper authorization, is unethical and often illegal.

Clearly, giving credit was an issue long before computers were developed. With or without computers, the applicable principles are the same; however, the widespread availability of computers has made definition of proper attribution more complicated and the principles easier to violate. For example, reports written for the U.S. government are often in the public domain. When quoting from such publications, do the same rules of ownership of intellectual property apply?

Today, ownership is not always clear. For example, software may be developed by a number of people, each making a substantive contribution to the development of the final product. Individuals may have difficulty determining what does belong to them, what to others, and to what degree. In some instances, proper use of property may also become an ethical issue.

Some of the questions that arose in discussing the scenarios on this controversial topic were:

- Under what circumstances can an individual's name or computer be used without seeking permission?

- Is inadvertent or careless use of a commercial software package unethical?

- What changes to a product are sufficient for it to be considered a new product?

- How should the rights to use and distribute property developed with a mix of individual and organizational resources be shared?

- Are there ethical (in addition to legal) considerations that should govern the distribution of property?

Several principles emerged from discussion of these and other questions. The availability of property belonging to others, including their names, phone lines, and computer systems, does not justify using or abusing it. In written works, all known sources should be accurately cited, both to credit the originator of the ideas and to assist readers in locating related studies. How the ideas are acquired — whether from a book, conference, computer forum, or informal discussion — is immaterial.

Unethical behavior requires intent. If individuals have an informed belief that their actions are right and other responsible people agree, they cannot be adjudged unethical. On the other hand, good intent does not justify taking what would otherwise be considered an unethical action. Professionals have a higher duty than others to exercise care in handling others' property and to be aware of relevant laws and contractual agreements. They are expected to understand what is ethical and what is not. This view does not imply that non-professionals are exempt from ethical and legal requirements, but merely that professionals should be judged more stringently when they stray from established ethical and legal guidelines.

While professionals are obligated to treat others ethically, they are also expected to act in their own best interests. This obligation includes declaring any proprietary interest in intellectual properties and refraining from providing so much detail that their ideas are easily appropriated.

Scholars and other professionals have the right to determine when their work is ready for distribution. Others should not unilaterally preempt that right. Rules for determining ownership of properties developed using varying degrees of employer and employee resources are lacking, however. Organizations need to work toward developing guidelines that can be used in avoiding or resolving conflicts that arise over ownership.

In questions of property distribution, distinctions must be made between private and public entities. Public entities, such as government agencies, are bound to distribute their resources equitably. Individuals and private organizations are not. Unless its charter, constitution, or bylaws state differently, an organization can keep or disburse its discretionary resources in whatever way its management chooses.

SCENARIO III.1

STUDENT, SERVICE DIRECTOR: EXPLOITING VULNERABILITIES IN A UNIVERSITY COMPUTER SERVICE

Without malicious intent, a computer hacker was scanning telephone numbers with his microcomputer and identifying those numbers that responded with a computer tone. He accessed one of these computers, using a telephone number he had acquired. Without entering any identification, he received a response welcoming him to an expensive and exclusive financial advisory service offered by a large bank. He was offered free of charge a sample use of some of the services if he would give his name and address. He provided someone else's name and address and used the free promotional services. This stimulated his interest in the services the bank charged for and gave him sufficient knowledge of access protocol to attempt to use the services without authorization. He gained access to and examined the menus of services offered and instructions for use. However, he did not use the services. By examining the logging audit file and checking with the impersonated customer, bank officials identified the computer hacker and claimed that he had used their services without authorization.

Hacker: Scanning Telephone Numbers for Computer Tone			
	Total	**Unethical**	**Not Unethical**	**No Ethics Issue**
General Vote	25	22	3	0
Subgroup Vote	6	6	0	0

Opinions

The entire subgroup and nearly all the general group believed scanning telephone numbers for computer response is unethical. At the very least, it creates a nuisance, invading the privacy and disturbing the peace of the holders of those telephone numbers. Curiosity does not justify this intrusion. In scanning, the hacker may also have interfered with others' legitimate use of the telephone.

Several group members questioned the absence of malice in the hacker's motives. ("Hacker" is not necessarily a pejorative term.) One remarked that what the hacker was doing was akin to testing doors of closed stores at night to see if any are open: Unless the person is a police or security officer on duty, his or her motivation in doing this would certainly be suspect. Another participant commented that calling someone's phone number is wrong except in a reasonable attempt to contact that person.

One participant asked whether the hacker was a professional and thus could be held to a standard of professional ethics. If the hacker was not a professional, he should be judged according to traditional concepts of right and wrong, moral and immoral. Applying these concepts to this case, the hacker acted immorally and, in some jurisdictions, perhaps even illegally.

The few who believed the hacker's action was not unethical granted that he was a nuisance, but said he was using publicly available access. This is not unethical if the hacker goes no further, that is, does not use the commercial service he discovers.

Hacker: Accessing a Computer System After Being "Invited" to Do So

	Total	Unethical	Not Unethical	No Ethics Issue
General Vote	25	12	13	0
Subgroup Vote	6	2	4	0

Opinions

A bare majority of the general group said the hacker's accessing the bank's computer system was not unethical under the circumstances; that is, the bank made no effort to prevent entry. The door was open with an invitation to enter. Simply providing a telephone number is a naive way to control access.

Nearly as many general group members said that accessing the system was still unethical. Having stumbled on the number in the process of scanning (itself unethical), the hacker was not truly invited to use it by a bank representative. The bank's failure to provide security does not excuse the hacker's behavior. This "invitation" was dishonestly acquired. The hacker also did not comply with the bank's conditions on the invitation — that he provide his name and address. He gave another's name and address, not his own.

Hacker: Using Someone Else's Name and Address

	Total	Unethical	Not Unethical	No Ethics Issue
General Vote	25	23	1	1
Subgroup Vote	6	6	0	0

Opinions

All but one group member agreed that using someone else's name and address was unethical. Using someone's name without permission constitutes theft, fraud, and misrepresentation. By impersonating another, the hacker potentially exposed that person to charges of unauthorized use of the computer. Some cited this act as clear evidence that the hacker knew that his previous actions, namely scanning telephone fines and accessing the computer system, were wrong. Otherwise, they contended, he would not hesitate to use his own name. The two dissenters gave no rationale for their positions.

Bank Officers: Claimed the Hacker Used Their Services Without Authorization

	Total	Unethical	Not Unethical	No Ethics Issue
Subgroup Vote	6	4	2	0

Opinion

The subgroup also considered the ethicality of the bank's contention that the hacker had used its services without authorization. Subgroup members decided 4 to 2 that the claim was unethical. As one remarked, the bank had no proof that the hacker had actually used the services, only that he had accessed them. The other side of the argument is that the bank had not provided the hacker with the number and that he had tried to cover his tracks by using another's name. Moreover, the logging audit file showed the hacker was in the system; thus, it would be logical to infer that he had used the services.

General Principles

One's name confers a right of ownership. Using another's name without permission is unethical.

The right of individual privacy holds calling someone's telephone number unless you are trying to communicate with that person is wrong.

Computer systems are property, just as an automobile, an office, and a house are property. They should be accorded the same respect.

(See also position paper in Appendix.)

SCENARIO III.2[5]

STUDENT, SERVICE DIRECTOR: EXPLOITING VULNERABILITIES IN A UNIVERSITY COMPUTER SERVICE

A manager in a computer facility quit his job and went into business as a consultant. He had been frustrated because he thought that his employer ignored his many suggestions for needed

---

[5]Scenario II.1 in the 1981 book.

improvements, including better security to protect data and programs stored in a large, multiaccess computer system. He had been given a secret password and was authorized to use it to gain access and use the computer services during his employment. At termination, the company did not tell him he was no longer authorized to use the services, and did not invalidate the password.

The former manager returned to the company, offering his consulting services to assist in improving computer security. The offer was refused. He then used the password from his own terminal and office telephone to extract copies of large amounts of data and programs for the sole purpose of presenting the material to the company to show them the computer operation was insecure.

Consultant: Misusing Outdated Access Authority to Demonstrate Computer Operation Insecurity				
	Total	Unethical	Not Unethical	No Ethics Issue
1987				
General Vote	24	23	1	0
Subgroup Vote	6	6	0	0
1977				
General Vote	28	18	0	1
Subgroup Vote	11	10	1	0

Opinions

The majority of the group members in both 1977 and 1987 agreed that the consultant's misuse of access was unethical; 1987 participants were nearly unanimous in this stance. Several of the 1987 general group commented that the consultant's action was comparable to someone leaving a job, keeping the office keys, and using the keys to return to his former employer's offices to remove copies of documents. They did not believe the consultant's intent, namely to show the weaknesses of the company's system security, justified his taking an action that was unethical and probably illegal. The one individual in the 1987 group who considered the consultant's action not unethical cited the consultant's purpose as justifying the action he took.

Security Officer: Failing to Invalidate Departing Employee's Password				
	Total	Unethical	Not Unethical	No Ethics Issue
General Vote	24	10	1	13
Subgroup Vote	6	5	1	0

Opinions

The 1987 group also considered the ethicality of the company security officer's failure to invalidate a departing employee's password. The subgroup majority believed the security officer acted unethically and incompetently; he made it easy for the consultant to break into the system. Although agreeing that the security officer was incompetent, a slight majority of the general group argued that unknowing incompetence is not an ethics issue. If the security officer's responsibilities did not include invalidating departing employees' passwords, then upper management was to blame for not establishing such a policy.

General Principles

1977

Two wrongs do not make a right.

Violation of trust can only be countenanced when the act in question is undertaken in the interest of the community.

The end does not justify the means.

1987

Nonemployees, whether previously employed by a company or not, should obtain explicit authorization to enter a company's premises either physically or electronically. Under any other circumstances, the absence of such authorization is implicit.

Employers have a responsibility to exercise due diligence in protecting their property.

SCENARIO III.3

MICROCOMPUTER USER: INADVERTENTLY CONVEYING COMMERCIAL SOFTWARE IN VIOLATION OF LICENSING AGREEMENTS

A microcomputer user purchased and legitimately used numerous commercial software packages protected by a typical license agreement. He wrote his own programs that called for the use of the commercial packages. Because his friends wanted copies of his programs, the programmer copied them and inadvertently, without noticing, also copied the commercial program onto diskettes. His friends proceeded to use his programs and, without knowing, also used the commercially available programs.

Programmer: Copying His Program Enhancement and Commercial Programs for His Friends				
	Total	Unethical	Not Unethical	No Ethics Issue
General Vote	24	8	6	10
Subgroup Vote	6	0	0	6

Opinions

The subgroup members concurred that no ethics issue was involved in the user inadvertently copying commercial programs; they reasoned that ethics requires a conscious decision to perform or avoid an action. People cannot be accidentally unethical.

Many in the general group agreed with the subgroup. These participants viewed the programmer's inadvertent act as human error — negligent, but not unethical. If the user had knowingly copied the programs, or if he were willfully negligent, he would be acting unethically. Some members maintained that if he were a professional computer programmer, his action would be negligent, and such behavior would be unethical.

Those who believed he did act unethically said that he failed to exercise due care in handling the property of others. He should have known what he was copying.

Friends: Using the Copied Programs, Including the Commercial Portions				
	Total	Unethical	Not Unethical	No Ethics Issue
General Vote	25	1	9	15
Subgroup Vote	6	0	0	6

Opinions

Here again, the subgroup concurred that no ethics issue was involved in using the copied programs. One participant noted that his friends might have assumed that the programmer had negotiated the right to distribute the programs. They had no way of knowing whether that was true, and no responsibility to know. The friends could also be ignorant of copyright laws; they might even think that the programs were in the public domain unless the licensing agreement was appended to the program package. For the same reasons, some group members voted that such use was not unethical.

One individual disagreed with the rest of the group, contending that the friends were violating property rights whether they knew it or not They did something wrong and that was unethical, although they might be less culpable than someone who did the same thing intentionally.

General Principles

Ethics requires intent. If the intent to do or not do something is not present, no ethical issue is involved. Responsibility requires knowledge in ethics, if not in law.

Professionals should exercise due care in handling the property of others.

Professionals are expected to know and abide by relevant laws and contractual agreements.

SCENARIO III.4[6]

CUSTOMER: USING SOFTWARE FIRM'S PROPOSAL TO SOLICIT COMPETITIVE BIDS

The president of a software firm proposed to develop an innovative accounts receivable computer program package with unique features that would be particularly valuable to his customer. The package was described in detail in the firm's proposal. The customer studied the proposal, then asked other firms to bid on the same package. Another firm submitted a substantially lower bid and won the contract.

---

[6]Scenario I.12 in the 1981 book.

Customer: Getting Competitive Bids on the Software Firm's Unsolicited Proposal

	Total	Unethical	Not Unethical	No Ethics Issue
1987				
General Vote	24	7	10	7
Subgroup Vote	7	2	4	1
1977				
General Vote	26	16	8	2
Subgroup Vote	12	4	7	1

Customer: Getting Competitive Bids on the Software Firm's Solicited Proposal

	Total	Unethical	Not Unethical	No Ethics Issue
1987				
General Vote	25	22	2	1
Subgroup Vote	7	2	4	1
1977				
General Vote				
Subgroup Vote	8	6	2	0

Opinions

If the proposal was unsolicited, most of the 1977 subgroup contended, the software firm president must expect to compete for the right to the contract. Barring a customer from selecting another contractor merely by sending an unsolicited proposal would be unfair. Therefore, the customer did not act unethically in seeking a lower bid.

If the proposal was solicited, however, the president had a right to assume that his proposal was confidential according to 1977 subgroup members. The customer would be unethical to violate that confidentiality by using the information in the proposal to generate a lower bid. A proposal represents an effort to produce new ideas or approaches. If a customer makes use of these ideas, they maintained, the originator should be compensated. The customer's failure to pay for the originator's effort and his undeserved profit in the form of a lower bid represented unfair exploitation or theft. No ethical issue would have been involved had the customer's solicitation stated that the proposal was to become his property, to do with as he wished.

The 1987 general group majority agreed with the 1977 subgroup majority that the recipient of an unsolicited proposal is not bound to maintain its confidentiality. In fact, in many situations, regulations may mandate competitive bids. For example, sole source solicitations are increasingly difficult to arrange in U.S. government contracts. Sole source proposals are typically approved only if a vendor has unique capabilities. U.S. government agencies often pay a contractor to develop the specifications to be contained in a request for proposal (RFP). Because the company that developed the specifications is thought to have an unfair advantage, it is not allowed to bid on the contract.

Those dissenting from the majority opinion believed the customer acted unethically — whether the proposal was solicited or not. One participant asserted that even if the proposal contained no formal declaration of proprietary interest, an informal contract of trust should be assumed, and informed consent should have been sought. Another said that the customer's action exploited the software firm.

General Principles

It is foolhardy to assume anything in bidding and contracting.

The rules for bidding on a contract should cover all cases and choices of action, and they should be known to all parties.

Individuals are responsible for protecting their own legitimate interests.

A formal declaration of proprietary interest should be included in any proposal.

An unsolicited proposal should not include so much detail that it becomes easy for someone else to do the work.

SCENARIO III.5

STUDENT, SERVICE DIRECTOR: EXPLOITING VULNERABILITIES IN A UNIVERSITY COMPUTER SERVICE

Searching for new product ideas, an independent commercial programmer purchased a highly popular copyrighted software package and studied it thoroughly. He concluded that he could produce a new package that would be faster, have greater capacity, and offer additional features. He also concluded that the market would be users of the commercial package that he had studied; his new product would replace the older one. The programmer realized that in some respects he could not improve the existing product and that compatibility between his product and the existing one would attract users and minimize the transition to his new product.

The programmer went ahead and developed the new product, meeting the higher performance and new feature capabilities that he had envisioned. The keyboard codes and screen formats (except for the additional features) for the new product were the same as those for the existing product. The computer program, however, was different and independently produced. The new manual was also entirely different from the existing product manual in content and style. The programmer gave his product a new name but advertised the value of its compatibility with the existing product.

The new product was highly successful. The company that produced the existing product, however, complained that the programmer had acted unethically in producing the new product. Although the company threatened criminal and civil legal action, it never followed through with litigation.

Programmer: Piggy-Backing on Existing Software Product Without Permission				
	Total	Unethical	Not Unethical	No Ethics Issue
General Vote	25	1	22	2
Subgroup Vote	5	0	5	0

Opinions

A large majority of the participants viewed the programmer's producing a new software package built on an existing product as not unethical. Such improvements, they argued, are at the heart of competition. The primary question is whether copyright laws were violated; most believed they were not. The programmer did not steal code or any other proprietary information. (Most did not consider the keyboard commands and screen formats proprietary.) The programmer's new software could be likened to reverse engineering, which the participants considered well within the bounds of fair competition. The fact that the company that owned the original product complained, but did not sue, seemed to bear out their contention that no copyright infringement occurred.

One group member disagreed, arguing that the programmer should have obtained permission to use the keyboard commands and screen formats and paid royalties on them. Two group members considered this case a legal, rather than an ethics, issue. As one stated, "This situation is no different than any other product innovation."

General Principles

Significant improvement of a software product results in a new product and should be treated as one. (In the case of unpatented, but possibly copyrighted software, the new program is assumed to be different and independently produced; that is, no proprietary or copyrighted property of another, such as source code, has been stolen.) This is true even when the function of the new product is similar to the original.[7]

Public formats, conventions, and standards are not copyrightable anymore than are the general shapes of other products such as cars or coffee makers.

Innovation must be encouraged for the good of society; therefore, individuals should be rewarded for being innovative.

(See also position paper in Appendix.)

SCENARIO III.6[8]

MANUFACTURER: IMPROVING ON A COMPETITOR'S PRODUCT USING THE COMPETITOR'S COPYRIGHTED MANUALS

A manufacturer designed and produced a product in the computer field. It was widely sold and successful. Manuals describing in detail how the product functioned and its characteristics were produced for customer organizations. The manuals were copyrighted and sold to the general public.

A competitor of the manufacturer, who was knowledgeable of the design of the product, decided to produce a competitive product. He bought copies of all the manuals and, based only on these manuals and his understanding of them, used new, faster, and

---

[7] A number of legal cases have recently dealt with the types of issues raised in this and the next scenario and appear to be in conflict with this principle.

[8] Scenario II.3 in the 1981 book.

less costly technology to design and produce an improved product with identical functions. He marketed the product in competition with the manufacturer, advertising that it was capable of performing the same functions and that the same manuals could be used.

Competitor: Marketing a Product Developed from Copyrighted Manuals

	Total	Unethical	Not Unethical	No Ethics Issue
1987				
General Vote	23	0	22	1
Subgroup Vote	6	0	5	1
1977				
General Vote				
Subgroup Vote	24	3	13	8

Opinions

Although group members were much closer to a consensus in 1987 than in 1977, participants still basically agreed that the manufacturer's competitor was not unethical. A few in the 1977 group considered the action unethical; none of the 1987 group did. The reasons for the unanimity were basically the same: Producing a better product that replaces an existing product falls within the realm of fair competition; the competitor did not steal code or any other proprietary information; the new product could be likened to a case of reverse engineering, which workshop participants considered well within the bounds of fair competition. The competitor did not act deceptively because he stated that customers must purchase the manuals from the manufacturer.

Copying the manufacturer's manuals instead of simply indicating that they were available from the other manufacturer and could be used for the new product would have been unethical. The manufacturer could have bundled the manuals into the cost of the product in a competitive counter move.

Those who said no ethics issue was involved used the same reasoning as the majority in reaching a slightly different, but compatible, conclusion. One group member indicated that this case was a legal, but not an ethics, issue saying, "This situation is no different than any other product innovation."

General Principles

Improvements to existing products should be encouraged within the constraints of fully honoring patent, trade secret, and copyright protections. This is the nature of fair competition and is not unethical.

Innovation must be encouraged for the good of society; therefore, individuals should be rewarded for being innovative.

SCENARIO III.7[9]

PROGRAMMER: TAKING PERSONAL PROGRAM TO A NEW POSITION

A computer programmer worked for a business enterprise that was highly dependent on its own computer system. He was the sole author of a computer program, and his manager was only nominally aware of its existence. He had written it and debugged it on his own time on a weekend, but had used his employer's materials, facilities, and computer services.

The programmer terminated his employment, giving due notice, and with no malice on his or his manager's part. He immediately went to work for a competitor of his former employer.

---

[9]Scenario II.7 in the 1981 book.

Without his former employer's permission, he took the only copy of the program with him to his new employer and used it in his work. He did not share it with any others. The new employer was not aware of the program or its use, but it enhanced the programmer's performance.

Programmer: Taking Personal Program to New Position

	Total	Unethical	Not Unethical	No Ethics Issue
1987				
General Vote	25	0	23	2
Subgroup Vote	8	0	8	0
1977				
General Vote	27	13	11	3
Subgroup Vote	10	5	5	0

Opinions

The participants' perception of taking personal computer programs to new jobs seems to have clarified over the last 10 years. Whereas in 1977 the groups were split over whether the programmer's action was ethical or not, in 1987 no one considered the action unethical. The 1987 group basically accepted the analogy of a mechanic who has his own tools (skills) and adjusts or calibrates them using his employer's equipment. The better his tools, the better the job he can do. Because the tools belong to him, using them in a new position is not unethical.

The mechanic's tools and the program represent the same situation. The program sharpened the programmer's skills and enabled him to perform more effectively. The company benefited from this while he worked for them. Because of his improved skills, he could get another job. He did no harm to his former employer and therefore his action was not unethical.

General Principles

Professionals possess the tools of their trade whether they are physical property or intellectual property.

SCENARIO III.8[10]

PROFESSOR, COMPUTER CENTER DIRECTOR: CLAIMING OWNERSHIP OF SOFTWARE DEVELOPED AT A UNIVERSITY

A professor of computer science at a university developed an advanced simulation language, a compiler for it, and a production system called SIMPAC.[11] The total cost of development was borne by him. He attempted to sell licensed use of SIMPAC to clients and treated it as a trade secret. The system was not flexible enough: He realized it was not sufficiently developed and documented and was not marketable. He and some of his colleagues used the system for their university-sponsored work on the university's computer. Over several years he improved the system at university expense, and the interest of others in it increased. However, he thought that the system was not yet sufficiently developed and refused requests from other universities for its use. He felt that when they tried to use it and had difficulty, they would reject it and it would reflect badly on his reputation.

The director of the university computer center also received requests from other universities for copies of SIMPAC. He was not aware of the professor's original investment in it, but knew the professor was reluctant to distribute copies. The director assumed

---

[10]Scenario II.2 in the 1981 book.

[11]A set of programs to translate the language into computer languate and execute a simulation on the computer.

it was owned by the university and sent copies of the system on tape, with out-of-date and limited documentation, to his friends at the other universities in exchange for favors. He did not tell the professor he had done this.

The professor discovered the director's action and complained, saying he would remove his system from the computer center. The director made a copy of the tape before the professor could take action, and claimed the university now had its own copy of the system, and it belonged to the university now, not the professor.

Professor: Refusing to Distribute Copies of SIMPAC and Not Obtaining Agreement About Ownership

	Total	Unethical	Not Unethical	No Ethics Issue
1987				
General Vote	26	0	21	5
Subgroup Vote	6	0	4	2
1977				
General Vote	16	5	10	1
Subgroup Vote	7	0	7	0

Opinions

In both 1977 and 1987, the majority of participants believed the professor's refusal to distribute copies was not unethical. He should not be expected to release a program until it was ready for use. The professor has that prerogative whether the program is his sole property, jointly held by him and the university, or the sole property of the university.

The opinions expressed in 1977 and 1987 differed in two respects. One is that several individuals in the 1987 group believed no ethics issue was involved. The other was that five individuals in 1977 believed the professor was unethical in refusing to distribute his program; no one in the 1987 group held that position. The 1977 minority reasoned that if the professor were receiving a salary while developing the program, the university had some claims on it, especially given the improvements made to the system at university expense.

Computer Center Director: Distributing the SIMPAC System				
	Total	Unethical	Not Unethical	No Ethics Issue
1987				
General Vote	25	24	1	0
Subgroup Vote	6	5	1	0
1977				
General Vote	22	17	2	3
Subgroup Vote	8	7	0	1

Opinions

A large majority of both the 1977 and 1987 groups considered the computer center director's action unethical. The 1977 group argued that the center director had an obligation to protect the professor's reputation. At the least, the director should have consulted with the professor to learn the reasons for his reluctance to release copies and verified actual ownership of the program. The director also acted unethically in using the professor's work to repay personal favors. This behavior remains an issue today.

Computer Center Direcctor: Taking Unilateral Action and Copying the Tapes (Not Voted on Separately in 1977)			
	Total	**Unethical**	**Not Unethical**	**No Ethics Issue**
General Vote	24	21	3	0
Subgroup Vote	6	6	0	0

Opinions

In determining that the computer center director acted unethically in copying the tapes, the entire 1987 subgroup and most of the general group contended that the computer center director had no right to take unilateral action, especially when ownership of the work had been brought into question. Those dissenting noted that the university did have a claim to the work, and the computer center should protect that claim. These same arguments were advanced by the 1977 participants.

The 1977 participants were split over whether the director acted properly to protect the university's interests. Some argued he had an obligation to safeguard computer center property and programs until the issue of ownership had been resolved. The majority did not feel that his copying the program was a wise or effective move. Both groups saw a need to develop clearcut means of determining ownership of properties developed using varying degrees of an employer's and an employee's resources.

General Principles

Software should not be sold to others until it has been properly developed, tested, and documented.

Scholars or other professionals have the right to determine when their work is ready for distribution. Programs developed by faculty should not be distributed without the author's consent.

SCENARIO III.9

COMPUTER COMPANY EXECUTIVE: CAUSING INEQUITY IN ACCESS TO COMPUTER TECHNOLOGY

A computer executive lived in a small suburban town with two grammar schools, one in an affluent neighborhood and the other in a poor neighborhood. The computer executive's children attended the school in the affluent neighborhood. Wishing to ensure that his children had a quality education, he influenced his company to provide microcomputers and computer programs for the classrooms, as well as computers to automate the administration of the grammar school in the affluent neighborhood. The students in that grammar school gained a high degree of computer literacy, which gave them a significant advantage over the students in the other school who had no access to computer products in their education.

Computer Executive: Influenced His Company to Provide Computers to His Children's School, but Not the School in the Poor Neighborhood				
	Total	Unethical	Not Unethical	No Ethics Issue
General Vote	26	6	13	7
Subgroup Vote	5	2	3	0

Opinions

The subgroup split in deciding whether the executive's actions were ethical. The majority argued that no one is obligated to be philanthropic, and, unlike a government agency, individuals and companies are not obligated to be equitable in distributing their resources. Being charitable is voluntary. Most general group members agreed with this opinion; private entities have the right to choose how they disperse their disposable resources.

Those who did not consider the executive's using his influence an ethics issue believed that ethics do not require the righting of all inequities. Although helping both schools might be best, helping one is better than none at all. On balance, therefore, the good effect outweighs any bad.

Some participants also pointed out that the computer executive, in his role as parent, was acting in his children's best interests, perhaps a higher obligation. For this reason those who perceived the executive's role as unethical said that he should not have involved himself in this decision at all — he put his own interests ahead of those of the community. He was wrong to use his position to gain an advantage for his family at others' expense. He should have been concerned about the social issue — widening the gap between the haves and the have nots. His action was even more unethical, according to the participants, if he did not disclose his self-interest to his company. One individual even equated charitable donations by a publicly held company to theft from its stockholders.

Computer Company: Providing Computers to One School and Not the Other

	Total	Unethical	Not Unethical	No Ethics Issue
General Vote	26	6	14	6
Subgroup Vote	5	2	3	0

Opinions

Both the subgroup and the general group voted similarly to the first issue raised for much the same reasons. Part of the discussion veered, however, to the company's motives. The donation might be viewed as providing a marketing advantage. It might also backfire; the company could be seen as elitist for helping only the school in the better area. Because the town had only two schools, favoring one over the other is bad public relations. The company would have been wiser to give the computers to the school system, which would then be ethically obligated to distribute them equitably.

Some participants thought the school system should ensure that access to computer resources is equalized between the two schools, even if the company specifically donated the equipment to one. This possibility presumes that one entity had jurisdiction over both schools. If true, that entity has a responsibility to provide the best education to all students.

General Principles

Private individuals and companies are not ethically bound to distribute their resources equitably. Public agencies, however, are so bound.

Equity and equal opportunity are ethical values imposed on persons in public positions who must be scrutinized for their distribution of resources, but not on persons acting as private citizens or representing companies who make charitable contributions.

SCENARIO III.10

CONSULTANT: USING SOFTWARE COPIED BY CLIENT

A consultant was formally told by his employer that copying any software product without the manufacturer's authorization was strictly forbidden.

In the course of his work on a project, the consultant was asked by the client to copy a copyrighted software product he needed to perform his tasks. The consultant told the client that this was forbidden by his employer and was not in conformance with his code of ethics. In front of the consultant, the client copied the software, handed it to him, and told him that now he could work. Knowing that the client was at fault, the consultant used the pirated software.

Client: Copying Software and Instructing Consultant to Proceed with Assignment				
	Total	Unethical	Not Unethical	No Ethics Issue
General Vote	25	25	0	0
Subgroup Vote	6	6	0	0

Opinions

The subgroup and the general group unanimously agreed that the client acted unethically. Copying the software was not only an unethical violation of another's property rights, but illegal as well. One participant noted that pressuring the consultant to use software that he knew was pirated was yet another unethical act.

Consultant: Using Pirated Software Provided by the Client				
	Total	Unethical	Not Unethical	No Ethics Issue
General Vote	25	25	0	0
Subgroup Vote	6	6	0	0

Opinions

Here again, both the subgroup and general group unanimously judged the consultant's action as unethical. The key is that he knew it was a pirated copy. If he had not known, his use of the software would not have been unethical. By using the software, the consultant became an accomplice of the client. The subgroup believed he should have held to his own code of ethics and the policy of his employer. The general group agreed, although one participant noted that a great deal depended on what he did later. For example, if the consultant purchased the software within a few days, a case might be made for using the copied software as an interim expedient.

Group members acknowledged the consultant's predicament and his probable fear of losing his client if he did not comply. However, they said his first loyalty should be to his own values and his employer's policies and reputation. If the software were required for the project, he should have asked his employer to acquire a legal copy.

General Principles

One should be willing to stand up for one's values regardless of the pressure applied.

Employees' loyalty requires that they observe the policies and protect the reputation of their employer.

The knowing use of illegally acquired property is as unethical as the act of illegally obtaining (in this case copying) it.

Individuals must be capable of knowing they are acting unethically in order for the action to be unethical.

SCENARIO III.11[12]

PRESIDENT, SOFTWARE FIRM:
SELLING THE SAME PRODUCT TO A SECOND CLIENT

The president of a small software firm negotiated a contract with a company to provide major modifications to the computer operating system used in that company's large computer. The operating system was supplied to the company by the computer manufacturer, without separate charges but under license.

When the work was completed, the president of the software firm sold a copy of the enhanced operating system to another customer without informing the first customer. The second customer was also licensed to use the same operating system as the first company.

The president of the software firm stated that there were no contractual provisions covering the ownership of the enhancements of the operating system and that he had the right to use them for other purposes.

Software Firm President: Selling the Same Product to a Second Client				
	Total	Unethical	Not Unethical	No Ethics Issue
1987				
General Vote	26	1	15	10
Subgroup Vote	6	2	2	2
1977				
General Vote	26	13	8	5
Subgroup Vote	10	8	2	0

---

[12]Scenario II.6 in the 1981 book.

Opinions

Opinions shifted shifted significantly between 1977 and 1987. Half the 1977 general group believed the president acted unethically; only one individual in the 1987 general group agreed. The 1977 group concluded that selling (without permission) any product whose development was paid for by another client was unethical. If the software firm itself had borne the development cost, the sale would not be unethical.

After extensive discussion among the 1987 group, more than half asserted that the president's act was not unethical. The basic issue, they reasoned, was the content of the contract. The first customer would have exclusive rights only if that privilege were explicitly stated in the contract. As long as he was not violating a contractual agreement, the president was within his rights to profit from his work, even if the first customer paid for all the development.

The 1987 group considered whether a software developer, being more knowledgeable about such issues, would have an obligation to point out conditions to a customer that should be in the contract to protect that individual's rights. Although this might be reasonable if the customer were naive, suggesting that one party to a corporate negotiation ask for conditions that would not be in his company's best interests to provide would be a foolish business practice. A person has a right to assume that someone empowered to negotiate for a company can act in that company's best interests. Some participants also noted that the president was in effect selling his expertise, not a product.

General Principles

Any conditions in a contract should be stipulated and agreed on during negotiations. Some conditions, although not stipulated, may be implicit and understood.

Individuals or companies have the right to sell their expertise to more than one buyer.

SCENARIO III.12

AUTHOR: MODIFYING AND COMMERCIALLY PUBLISHING A PUBLIC GOVERNMENT REPORT

An author wrote a large report for a government agency and was properly paid for his services. The author had used a word-processing and desktop publishing system to write the report. With very little effort, he used the word replacement feature of the system to make slight changes in the report. Then he published the report as a commercial professional book without referencing the original report or crediting the government agency for its sponsorship.

Author: Revising His Report to a Government Agency Without Giving Credit for Sponsorship				
	Total	Unethical	Not Unethical	No Ethics Issue
General Vote	26	25	0	1
Subgroup Vote	6	6	0	0

Opinions

Nearly everyone agreed that the author was unethical in his failure both to reference the original report and to credit the government agency for its sponsorship. Some contractual or other legal requirements for the reference may exist, but if not, custom, scholarship, and courtesy (to the readers as well as the sponsoring agency) dictate that proper credit be given.

Several individuals indicated that the extent of the changes made to the document affected the importance of referencing the previous work. If the textual changes were extensive (instead of "slight") and the artwork redrawn, the unethical nature of the action would be less clear.

The one dissenter argued that the author was under no obligation to credit the agency for its support because the support was only financial, not creative. Thus, the omission was discourteous, but not unethical.

General Principles

It is dishonest and unprofessional to misrepresent derivative work as original. Complete references should always be provided.

Authors of scholarly works have an obligation to readers to help them find related studies.

It is better to err in overcrediting than to fail to give credit where it is due.

SCENARIO III.13

AUTHOR: USING GOVERNMENT REPORTS

An author read a government report and a book that was nearly identical to the report, both having been written by the same person. The author wrote a new book on the same subject and quoted verbatim more than 50 pages from the government report. He cited the source as the government report; however, he did not name the government report author nor did he acknowledge that the material was also published as a commercial book protected by copyright.

Author: Failing to Give Credit to Author of Government Report				
	Total	Unethical	Not Unethical	No Ethics Issue
General Vote	22	9	11	2
Subgroup Vote	6	5	1	0

Opinions

All but one member of the subgroup believed the author acted unethically, but the general group's opinion of the failure was nearly evenly split. A slight majority argued that the author did not act unethically for several reasons. First, a government report is in the public domain; anyone has the right to use material from such documents. Second, the author did cite the government report as the source of the material he quoted. Third, government reports often do not contain the name of the author, therefore, his citation of the agency as the source may have been complete. Fourth, even when the author's name is included in a government report, omitting it from a citation of the work is not uncommon. This author should not be held to a higher standard than ordinary scholarship requires. Presumably the agency paid for the work; the material is now the property of the agency and, by extension, of the public at large. Because the author made no attempt to conceal the source of his work, his action was not unethical.

Those who believed the author did act unethically suggested that giving less than full credit was suspicious, that it amounted to misleading the reader into thinking the work was original. Failing to give proper credit is unfair to the other author. A professional should strive to name all sources and to provide full references to other related work for the benefit of the readers.

Author: Failing to Acknowledge Commercial Version of the Government Report				
	Total	Unethical	Not Unethical	No Ethics Issue
General Vote	25	2	23	0
Subgroup Vote	6	1	5	0

Opinions

Both groups nearly unanimously voted that the author did not act unethically in omitting the commercial report citation from his book. The author had not quoted from the commercial book and was under no obligation to advertise someone else's product. Still, several members of the group suggested that he should have cited all known sources.

Those dissenting from the majority view argued that giving less than full credit is suspicious. Scholars have a responsibility to give others credit for their work. If the author had not been aware of the commercial version, obviously he could not be considered to be acting unethically.

General Principles

All known sources should be cited accurately to assist readers in finding related studies. Quoted material should be fully referenced. Persons who publish scholarly works should be given full credit for that work.

SCENARIO III.14

COMPUTER FORUM PARTICIPANT: USING ANOTHER PERSON'S IDEAS

A small group of scientists from across the country, while at a convention, decided to set up an ongoing computer forum on the topic of their common specialy. One of the scientists set up the system, and others tied in. They exchanged ideas and discussed the results of research still in the formative stage. Several months later, just as one scientist was about to send a manuscript to a journal for review, he noticed an article, just published, by a participant in the forum. As the scientist read the article, he realized that it contained many of the ideas in the paper he was about to submit for publication.

Forum Participant: Publishing Material Based on Other's Ideas				
	Total	Unethical	Not Unethical	No Ethics Issue
General Vote	25	22	3	0
Subgroup Vote	7	7	0	0

Opinions

Almost everyone assumed that the journal article did not give proper credit to the other forum participant for his ideas. Giving due credit is more than a matter of courtesy. They maintained that ideas should always be credited, regardless of how they are acquired, e.g., in a book, at a conference. Therefore, the entire subgroup and the majority of the general group concluded that the article writer's action was unethical.

Two of the dissenters argued that issues raised during open conversations, whether in person or through a computer forum, are in the public domain. Another group member assumed that the article did contain proper attribution.

General Principles

Scientists should acknowledge the source of their ideas, regardless of the circumstances under which they were acquired.

IV
Confidentiality of Information and Personal Privacy

Accompanying the growth of data bases in private and public institutions is a concentration of personal and other sensitive data that should be considered privileged and confidential. Because organizations are increasingly automating the processing of personal information without the consent or knowledge of the individuals affected, the opportunities to breach more people's privacy at less cost, to greater advantage, occur much more frequently today than a generation ago. Nevertheless, disseminating personal information presents the same ethical considerations today as it did before computers emerged. The secondary effects that encourage increased sharing of personal information such as reduced cost collection opportunities, and ease of distribution or access do not change the ethicality of breaching an individual's privacy.

Those who address the ethical issues surrounding sensitive data must be aware of the dilemmas posed by the various laws, regulations, and policies dealing with such data. On the one hand, sensitive data must be protected from disclosure to unauthorized individuals and organizations. Certain federal statutes thus prescribe legal penalties for unlawfully accessing or disclosing information (e.g., Privacy Act of 1974), prohibit disclosure but provide no criminal penalties (e.g., Right to Financial Privacy Act), and require safe-guarding of information (e.g., Tax Reform Act of 1976 and the Computer Security Act of 1987).

On the other hand, some statutes require disclosure of certain information to law enforcement agencies, to the individuals affected, and to the public (e.g., the Freedom of Information Act). At the federal level, many of the same statutes that mandate privacy have several exceptions allowing access for law enforcement purposes (e.g., the Electronic Communications Privacy Act of 1986). Moreover, nearly every state now has at least one statute providing for confidentiality of one or more categories of computerizable information.

Some of the critical questions that need to be addressed when confidentiality and personal privacy are at stake include the following:

- Should a researcher use data gathered for one study in another study without the consent of the data's custodian or the subjects involved?

- What are the implications of a student making an offensive questionnaire available to other students with access to a university computer?

- Is it ethically acceptable for a project manager to design a system that secretly conveys negative information about customers?

- Under what conditions should the manager of a computer service company release information on a subscriber's use of its services?

The group of scenarios presented in this chapter elicited strong reactions. Participants asserted that information should be used only for the purposes for which it was intended. In research, informed consent is required and human subjects have the right to withdraw from an experiment at any time. All research on humans should be halted if any harmful effects are noted.
People have a right to privacy except when they knowingly give up that right. Even in those cases, some situations warrant limitations on the information that is released and to whom it is given. Companies whose records contain data on their employees'

personal habits have a duty to protect that information to the extent permitted by law.

SCENARIO IV.1[13]

PROGRAMMER, MANAGEMENT, GOVERNMENT AGENCY: REVEALING A SYSTEM'S INADEQUATE PROTECTION OF PRIVACY

A computer programmer working for a state department of health assisted in the development of a computer application system. The system had few controls over access by unauthorized persons to data processed and stored by the system, and no monitoring or logging of access was performed. The data consisted of medical information on identified people.

The state had no laws controlling the confidentiality of the medical data. However, the agency had rules governing confidentiality and prohibited employees from unauthorized access and from taking copies of data. The rules were invoked only on a manual basis, and little was done to enforce them or monitor for violations.

The programmer thought that the computer system did not adequately enforce the rules of confidentiality, a situation that resulted in increased and dangerous exposure of the data. He brought this to the attention of his immediate manager and finally of higher management with no success in getting authorization to correct the deficiency. He was finally threatened with sanctions if he did not stop his efforts.

The programmer violated the rules of confidentiality and easily obtained a copy of medical data about himself from the system. He presented the copy of these data to a state legislator representing his legislative district. With the programmer's permission, the legislator revealed this action and the data at a public hearing of a legislative committee and the revelation was widely publicized in newspapers. This greatly embarrassed the agency, and the programmer was fired from his job.

---

[13]Scenario III.3 in the 1981 book.

Programmer: Disclosing a Computer System Vulnerability to an Outside Authority

	Total	Unethical	Not Unethical	No Ethics Issue
1987				
General Vote	23	1	22	0
Subgroup Vote	6	1	5	0
1977				
General Vote	28	5	22	1
Subgroup Vote	11	4	7	0

Opinions

The 1977 and 1987 group members agreed that disclosing vulnerabilities to outsiders is not unethical, with the 1987 group almost unanimous in their opinion. In 1977, participants argued that the programmer did no harm to others whose medical data were stored in the system; however, in breaking agency rules, he had to expect to pay the price for his courageous act. All but one subgroup member believed the course of action he took was the only honorable one open to him under the circumstances because he had exhausted the agency's normal administrative channels without success. As a citizen, he was free to take his case to his representatives. Because the legislature had oversight responsibility for the agency, the legislator might reasonably be considered the next appropriate level of authority.

The basic moral issue, according to the participants, is between the obligation to obey agency rules and the obligation to protect the public from invasion of privacy. The latter has a higher moral priority. Assuming he had taken all reasonable action within the agency before going to a higher authority, the programmer performed a highly ethical act in violating confidentiality only in order to preserve it, and only with his own data. His act was justified by society's gain. Violating an agency rule to demonstrate that it was inadequately enforced, if unethical, was certainly much less unethical than failure to act. Therefore, the net result was ethical.

The 1977 group made other points in defense of the programmer. Although the scenario states that he violated the rules of confidentiality, gaining access to information about himself cannot reasonably be construed as violating confidentiality. Rather, the question of ethics should concern his improperly gaining access to the information in order to prove his point. Because he was partly responsible for development of the system, he had a higher obligation to assure that the system properly protected the confidentiality of the information it contained. Nevertheless, to avoid any culpability as a disloyal employee, the programmer should have announced his intention to inform the legislator and terminated his employment. He could then have taken the evidence to the legislator free of any constraints as an employee.

A minority of the subgroup contended the two acts — taking the data to the legislator, and obtaining the data by compromising the system — should be separately evaluated. Informing the legislator was in accord with the programmer's professional duty to identify substandard situations and, if ignored, to seek outside assistance. Yet, he had no right to violate the very rules he sought to preserve. That act was unethical. However, the subgroup was divided on whether the programmer broke a rule, half of them maintaining that a rule that is not enforced de facto does not exist.

Generally, the 1987 participants believed the programmer's behavior was a justified case of whistle blowing. He had exhausted his options within the organization and had no choice but to seek recourse outside. Only one individual disagreed, arguing that his use of the data, even data about himself, was unauthorized and therefore unethical. The end can never be said to justify the means.

Agency Management: Refusing to Correct the Deficiency and Firing the Programmer

	Total	Unethical	Not Unethical	No Ethics Issue
1987				
General Vote	25	25	0	0
Subgroup Vote	6	6	0	0
1977				
General Vote	17	14	3	8
Subgroup Vote	7	7	0	0

Opinions

According to most 1977 participants, although not correcting the deficiency and firing the programmer were both unethical, the agency clearly had the right to fire the programmer for rule violation. Moreover, the agency had acted ethically in establishing the rules to protect privacy, and even though enforcement was lax, the agency was justified in expecting employees to abide by them. In most states, however, civil service regulations permit firing of employees only under the severest of restrictions. Practically, his firing would probably not withstand the publicity and resulting pressure from the legislature.

The 1977 majority also believed management acted irresponsibly in threatening the programmer with sanctions and failing to correct the problems. Retaliation and coercion of conscientious public employees violate public responsibility.

All 1987 participants concurred that agency management acted unethically. In not correcting the situation the programmer brought to their attention and in firing him, the managers failed in an important responsibility. Retaliating against someone acting in the best interests of the public is clearly unethical. One workshop participant even suggested that he should have been promoted. Others conjectured that the agency officials refused to correct the deficiencies in an attempt to disguise their own deficiencies and negligence, and fired the programmer in reaction to their embarrassment at being exposed.

General Principles

It is right to expose inadequate protection of confidentiality if the personal rights of others are safeguarded in the process, and the proper internal channels for protest and disclosure of the inadequacy have been exhausted.

Acts of civil disobedience may be illegal or may violate an agency rule, but under some conditions they may also be highly ethical. Violation of a law or rule cannot be equated with unethical behavior.

Each person must decide for himself how far to take a point of principle, assuming the action is not itself illegal.

Civic obligation takes precedence over loyalty to an organization.

Sometimes, one must incur personal injury in order to accomplish a higher good to society. Unfortunately, one cannot always expect society to compensate for any damages.

Acting on high personal principle may not be perceived as ethical by society, even though it is an act of conscience.

The right of employees to break rules on the basis of their personal and subjective judgment cannot be condoned, except when all other avenues have been exhausted because it would cause anarchy. Government could not survive.

Acting ethically is sometimes neither clear-cut, nor easy.

Having the legal right to take an action does not necessarily mean that the action is ethical.

SCENARIO IV.2[14]

SCIENTIST, PROGRAMMER: MAKING NEW USE OF PERSONAL DATA

A scientist employed by a university as a researcher learned that two different kinds of data on essentially the same subject pool were contained in two files stored in the university's computer. He believed that there would be significant scientific value in merging the files and reanalyzing the data.

Although the subjects' informed consent had been obtained for the earlier studies (they were students who had since graduated), their permission for this new use for the data had not been sought. The scientist was aware that it would have been desirable to seek permission of the subjects, but he decided not to do so, because it would have been time consuming and would have added considerably to the cost of the study he was proposing.

He thus asked one of the university's programmers to access the data, merge the files on the same subjects, and analyze the data as he indicated. The programmer did as the scientist requested.

Scientist: Merging Files Without Permission of Subjects				
	Total	Unethical	Not Unethical	No Ethics Issue
1987				
General Vote	24	23	0	1
Subgroup Vote	6	6	0	0
1977				
General Vote	28	19	8	1
Subgroup Vote	10	5	5	0

[14]Scenario III.4 in the 1981 book.

Opinions

The majority of both the 1977 and 1987 groups considered the scientist's action to be unethical. Unlike in 1987, when the group members were almost unanimous in their views, many 1977 participants argued that the scientist's action was borderline in terms of ethics. On the one hand, he certainly acted unethically by not getting the informed consent of the subjects, and by not consulting the original researchers about whether combining the two subject populations could invalidate the new research. On the other hand, the principle of informed consent has a practical limit. When data are collected about identifiable individuals, questions of informed consent must be addressed, but cannot be answered by generalizations. It would be a mistake to impose a blanket restriction on the use of personal information. A problem should be identified before restriction of use is imposed. One participant, in fact, believed the use and dissemination of data should not be restricted unless individuals in the group studied can be identified.

The 1977 group asserted that although informed consent is important, an independent committee should be consulted to weigh the benefits of the research against the costs when the costs of obtaining informed consent are prohibitive. Taking an independent action, without checking with those who collected the original data, appears to be unethical.

The 1987 groups came to essentially the same conclusions. They too believed the researchers in charge of the original studies and/or the university's human subject review committee should have been consulted. Using the data would have been less objectionable to them if the scientist had consulted these other sources, if he had removed any identification of the subjects (if getting informed consent were not feasible), and if use of the data would not harm the students. One member differentiated between experiments and analysis, maintaining that informed consent is required for the former, but not necessarily for the latter.

Programmer: Merging Files as Requested				
	Total	Unethical	Not Unethical	No Ethics Issue
1987				
General Vote	23	4	12	7
Subgroup Vote	6	1	2	3
1977				
General Vote	25	14	5	6
Subgroup Vote	9	6	1	2

Opinions

The 1977 participants were split over whether the programmer behaved unethically. However, a large majority of the subgroup and a slim majority of the total group believed he should not have merged the files without the consent of the researchers who owned the files, that is, those who had obtained consent from the subjects for the original research.

The opinions of the 1987 groups were the reverse of the findings of the 1977 groups. Those who said that merging the files was not unethical or not an ethics issue based their opinion on the programmer's not having any responsibility to question the scientist's right to access and merge the files. As long as the programmer did not have to enter the files surreptitiously — that is, breach the university's computer security to retrieve and merge the data — he had no reason to question the scientist's request. It would be an institutional nightmare for subordinates to question each file merge and require approvals before performing one. A few group members disagreed, however, insisting that the programmer should have satisfied himself that the merging was proper before agreeing to do it. Although many universities have a human subjects committee, educational institutions and other organizations should establish policies to handle situations similar to the one in this scenario.

General Principles

A programmer or systems analyst should always seek direct and positive authorization for unprecedented use of data files from whomever he has identified, in his best efforts, as the custodian of the files.

Today, with more and more data bases, governments are merging files with increasing frequency, sometimes justifying the merges using specious legal arguments. Expediency, however, does not excuse unethical behavior.

Information should be used only for the purposes for which it was intended, unless permission for a different use of the data is obtained from the appropriate authority.

SCENARIO IV.3

ADMINISTRATIVE PERSON: BROWSING THROUGH PERSONNEL RECORDS

An administrative person working in his assigned area of accounts receivable found, as a result of browsing various menu screens, that he was able to access the personnel data base. He was not authorized to work on personnel applications, but the data base was not marked confidential.

He found various records of other employees including earnings, salary increases, date of birth, date employed, and other personal information. He confronted his immediate manager and stated that he was not receiving salary increases as fast as other people, claiming he "knew what was going on!" The manager replied that a salary increase was pending and the employee was being treated fairly.

Administrative Person: Unauthorized Browsing			
	Total	Unethical	Not Unethical	No Ethics Issue
General Vote	24	22	2	0
Subgroup Vote	6	6	0	0

Opinions

Nearly everyone agreed that the administrative person acted unethically when he browsed through files that were not part of his job duties and are commonly understood to be confidential. By examining personnel files, he violated his fellow employees' right to privacy with no legitimate need to know. The fact that others do it does not justify an unethical action. Many group members added that company management is also at fault for not providing greater security, or at least issuing a warning that the files are confidential, with access limited to authorized persons. A minority concluded that the absence of such security measures in effect authorizes access.

Administrative Person: Using Improperly Obtained Information to Negotiate a Raise			
	Total	Unethical	Not Unethical	No Ethics Issue
General Vote	24	22	2	0
Subgroup Vote	6	6	0	0

Opinions

Again, almost everyone asserted that using improperly obtained information about others for one's own ends is clearly unethical.

Several commented that such an act was also stupid. Many employers would consider it grounds for terminating employment. One participant argued, however, that although the browsing was unethical, once the administrative person had the information it was not unethical to use it.

General Principles

Browsing through information not clearly in the public domain and that one has not been authorized to access, especially personal data, is unethical.

Use of data obtained unethically is also unethical.

Others' rights to privacy must be protected.

Personnel files in a company are confidential except when necessary for disclosure in the public interest and properly authorized by a public entity.

SCENARIO IV.4

POLITICAL SCIENTIST, EXPERT: QUOTING STATEMENTS MADE ON A BBS

A political scientist engaged in a bulletin board system discussion with her peers on a controversial topic. Another expert on the subject contributed to the exchange in strong and decisive language in response to provocations by the political scientist. During an interview with a journalist on the subject, the political scientist quoted what the expert had said in the bulletin board system because it happened to be on her screen when the journalist called. The expert was chagrined to find that his strong words were quoted with attribution and that they were taken out of the context of responses to what others had said in the bulletin board system.

Political Scientist: Quoting Expert's Statement on Electronic Bulletin Board to Press			
	Total	Unethical	Not Unethical	No Ethics Issue
General Vote	25	0	25	0
Subgroup Vote	5	0	5	0

Opinions

Both groups decided unanimously that quoting a statement made on an electronic bulletin board system (BBS) with attribution was not unethical. They reasoned that an electronic bulletin board, like a regular bulletin board, is public and available to anyone who sees it. They compared posting messages on a BBS to writing letters to a newspaper editor or sending a postcard rather than a letter in an envelope. Like electronic mail, the letter is directed to one or more specific individuals with the reasonable expectation that only the individual(s) addressed will read the contents.

Prudent individuals should assume that a BBS is public, unless access is controlled, or unless or organization sponsoring the BBS gets the participants' agreement that the information exchanged will be held private and confidential. Even under these circumstances, BBS users should realize that individuals unknown to them and to the manager of the BBS may gain access and eavesdrop on what the participants believe to be private conversations. The group did not address the ethical nature of individuals engaging in such eavesdropping.

If one were to disagree with the majority opinion, one would have to believe the BBS to be a private, not a public, communication. To quote material gleaned from it without permission would then be an invasion of privacy and unethical.

Political Scientist: Quoting Expert Out of Context After Provoking a Strong Response				
	Total	Unethical	Not Unethical	No Ethics Issue
General Vote	25	15	10	0
Subgroup Vote	5	5	0	0

Opinions

The entire subgroup and the majority of the general group agreed that the political scientist was acting unethically. According to them, the key issue here was the intent of the political scientist, which they believed to be malicious. It seemed likely that she purposely, rather than inadvertently, quoted the expert out of context. They added that, as a matter of courtesy, she should have asked the expert's permission before quoting him to the press.

In disagreeing, the minority made the point that context is subjective. What one person may consider to be in context, another would say is out of context. Furthermore, people are often quoted out of context; an expert should be used to it.

General Principles

Statements made on closed bulletin boards, especially on controversial topics, should be assumed to be off the record and should not be quoted before permission is obtained from the source.

Electronic mail directed to one or more specific individuals using public services should be treated as privileged in the same manner as first-class U.S. mail.

Open bulletin boards are public, like physical bulletin boards, letters to the editor, or postcards. Closed bulletin boards or conferences are private.

To act with malice is unethical.

Acting with courtesy, although not a true ethics issue, helps ensure ethical behavior.

Experts should expect to be quoted out of context and, therefore, be circumspect in their pronouncements.

SCENARIO IV.5

PROJECT MANAGER: DEVELOPING SMART CARD THAT STORES CREDIT-WORTHINESS DATA UNKNOWN TO CARDHOLDERS

The manager of a smart card development project designed a new debit card system for a financial institution. One feature of the system allowed the institution to store in the card, without the cardholder's knowledge, creditworthiness information about the cardholder. As a result, many cardholders used their cards for transactions not knowing that negative credit information was conveyed that allowed the merchants to make immediate decisions concerning the size, nature, and conditions of the transactions.

Smart Card Developer: Designing Smart Card System That Permits Merchants to Secretly Determine Creditworthiness of Cardholder

	Total	Unethical	Not Unethical	No Ethics Issue
1987 General Vote Subgroup Vote	24	22	2	0
(secretly)	7	5	2	0
(openly)	7	1	4	2

Opinions

All but two members of both groups considered the use of a smart card containing information on creditworthiness unethical if the cardholders are not made aware that the information is being provided. Particularly unjustified would be supplying detailed transactional information about the buyer's purchasing patterns and other economic characteristics. Providing such information without telling the cardholders is an invasion of privacy. Moreover, members contended, users have a right to know what data are included on the card. If they know, then by accepting the card they in effect give informed consent.

One of the two who saw nothing unethical in this situation argued that credit history is no more an individual's property than medical history. Current credit and debit cards do not indicate the contents of the magnetic stripe. That would be poor security practice. People have certain legal rights to inspect and correct their credit history; beyond that, however, no one is forced to enter into such an arrangement with a financial institution. By doing so, the customer agrees to that institution's providing the needed information through the card instead of a telephone call to a credit agency.

One majority group member suggested that disclosure does not mean that the cardholder understands exactly how the card works. He thought the most appropriate use of the card would be similar to today's dial-up system. The card would track holders' purchases against their credit limits and notify the merchant only that the sale is or is not approved.

General Principles

People have a right to privacy unless they give up that right in a specific commercial transaction (e.g., accepting and using a credit card). Even under those circumstances, there should be limitations on what information is released and to whom.

People have a right to inspect and correct information in their credit records.

(See also position paper in Appendix.)

SCENARIO IV.6

PROGRAMMER: DEVELOPING MARKETING PROFILES FROM PUBLIC INFORMATION

An enterprising programmer used publicly available information stored in a variety of places or available by purchase from the Department of Motor Vehicles, mail order firms, and other sources to compile "profiles" of people (shopping habits, likely income level, whether the family was likely to have children, etc.). He sold the profiles to companies interested in marketing specialized products to niche markets. Some of his profiles were inaccurate, and the families received a large volume of unsolicited, irrelevant mail and telephone solicitations. They did not know why this increase in their junk mail and calls had occurred and found it annoying and bothersome. Other profiles were accurate, and families benefitted from receiving the sales materials.

Programmer: Developing and Marketing Sometimes Invalid Consumer Profiles				
	Total	Unethical	Not Unethical	No Ethics Issue
General Vote	26	8	15	3
Subgroup Vote	6	4	2	0

Opinions

Members in the general group and the subgroup came to opposite conclusions about the ethicality of selling consumer profiles compiled from publicly available sources. Although no one liked the practice, and some had personal experience with the annoyance of junk mail, the general group majority could find nothing unethical in the programmer's action. (One offered the caveat that it was not unethical provided the programmer did not represent his information as more accurate than it was.) They assumed that all information used was public (and thus involved no invasion of privacy) and not gathered by illegal or surreptitious means. Selling mailing lists and membership lists for use in direct mail and telemarketing is common practice. No harm is done; if people receive mail they do not want, they can simply throw it away.

Those who decided no ethics issue was involved essentially agreed with the majority. (They also assumed the profiles were based on public information.) As one member said, while the action was annoying and sleazy, it was not ethically impermissible. Another focused on the quality of the profiles produced, contending that ignorance of one's incompetence is not an ethical issue either. However, both believed that standards of professional ethics require a certain degree of competency. The minority view was that compiling these profiles was an invasion of privacy.

Unethical acts require intent If the programmer did not know that his profiles were inaccurate, and was therefore unintentionally causing his clients to waste money and those profiled to have overflowing mailboxes, he would not be acting unethically.

General Principles

Use of the computer as a marketing tool is not, per se, unethical.

Using personal information for purposes other than originally intended is unethical.

Use of publicly available, legitimately acquired information is ethical.

SCENARIO IV.7

INSTRUCTOR: USING STUDENTS AS SUBJECTS OF EXPERIMENTS

An instructor of a logic course decided to test a computer-assisted instruction (CAI) system under development. The large class was divided randomly into two groups. The instructor arranged a controlled experiment in which one group was taught in the traditional manner with a textbook, lectures, and tests, but with no CAI. The other group used the same textbook, lectures, and tests, but in addition used CAI. The grading practices were the same for both groups.

By the middle of the term, the instructor realized that the students in the experimental group who had access to CAI were doing much better than the students in the control group. Some students in the control group sensed this difference and complained that, although they paid the same tuition, they were being denied an educational opportunity offered to others. These students insisted that the instructor discontinue the experiment and allow them to use the CAI package for the remainder of the term. The instructor refused the students' request on the grounds that ending the experiment prematurely would vitiate the results of the experiment. The instructor pointed out that only by chance were they in the control group and, because free inquiry and research are the nature of the academic world, students should take part willingly in such experiments for the sake of advancing knowledge. At the end of the term, the grades in the experimental group were significantly higher than the grades in the control group.

Instructor: Using Students as Subjects of Experiments				
	Total	Unethical	Not Unethical	No Ethics Issue
General Vote	27	21	4	2
Subgroup Vote	8	7	1	0

Opinions

Both groups overwhelmingly asserted that the instructor acted unethically. First, the experiment harmed the subjects in the control group. Their grades were significantly lower than those in the group using CAI, a situation that could affect their chances of getting or maintaining a scholarship or entering the graduate school of their choice. Second, the instructor did not get the students' informed consent. Such consent is a normal requirement for any research experiment involving human subjects.

The minority who voted that using the students in an experiment was not unethical did not accept the contention that the students were harmed. They suggested that using two different classes rather than dividing one class into two parts would have been wiser, but pilot testing a new tool or procedure to see if it works before using it on the whole student body is only good pedagogical practice. Another argued that it wasn't really an experiment in the psychological sense.

Instructor: Refusing Student's Request to Discontinue the Experiment			
	Total	Unethical	Not Unethical	No Ethics Issue
General Vote	26	23	3	0
Subgroup Vote	8	8	0	0

Opinions

An even larger majority believed strongly that the instructor should have terminated the experiment when the students complained. One group member even argued that he should have been fired for refusing. Once the instructor determined that the control group was placed at a disadvantage, he should have given them an opportunity to recover. Even if they had initially consented to participate, the instructor acted unethically when he did not honor their request to discontinue the experiment. If the instructor had taken steps to mitigate the grading inequities, some members of the group might have voted differently.

The few who considered the instructor's action not unethical contended that students take their chances with different instructors who use different books (or even the same books) for the same course. It might be argued that last year's class, for example, which did not have access to CAI, was also harmed. It's an imperfect world and likely to remain so.

General Principles

The informed consent of human subjects is required before they can be used for any research experiment.

When deleterious effects are seen, the experiment should be stopped. To the extent possible, these negative effects should be reversed.

Human subjects have the right to withdraw from any experiment regardless of whether they initially gave informed consent.

SCENARIO IV.8

UNIVERSITY STUDENT: OFFERING LIMITED ACCESS TO A PORNOGRAPHIC QUESTIONNAIRE

Students at a university were each given small personal computer accounts on the university-owned mainframe. Through the student chapter of the Association for Computing Machinery (ACM), a forum system was set up. Any student with an account could sign on, read what had been entered into the forum, and add his or her comments. A discussion of sexual behavior developed. One student briefly described a pornographic questionnaire that had been distributed among students. The questionnaire asked in graphic detail whether the individual would or would not do certain things on a first date. The student also announced that he had put the questionnaire in one of his files and had authorized access under a particular sign-on ID. He revealed the sign-on ID only to those who wanted to see the questionnaire. He warned those who might be offended that the questionnaire was crude. Several weeks later, the student was called into the Dean of Students' office and threatened with expulsion. The university had heard about the questionnaire and had traced it to him through his comments in the forum.

Student: Making a Pornographic Questionnaire Available to Other Students

	Total	Unethical	Not Unethical	No Ethics Issue
General Vote	25	1	21	3
Subgroup Vote	7	0	7	0

Opinions

Most group believed providing the questionnaire to other students was not unethical. Although the questionnaire might be in bad taste, no one was forced to read it. The student clearly identified its contents and warned that some people might be offended. Those accessing it did so with informed consent, not accidentally or under any duress, and need not submit their answers to anyone, or even complete the questionnaire at all. The one dissenter contended that those answering the questionnaire were being exploited. Unless the student or some other party made use of completed questionnaires, group members did not view the questionnaire's availability as exploiting anyone.

Dean of Students: Threatening the Student with Expulsion

	Total	Unethical	Not Unethical	No Ethics Issue
General Vote	26	17	2	7
Subgroup Vote	7	7	0	0

Opinions

A clear majority concurred that unless the student behaved unethically or violated rules prohibiting such use of the computer, the Dean of Students acted unethically in threatening him for his action. Under the circumstances, the dean abused his power and violated the student's right to freedom of speech. Those participants who believed the threat was not unethical or not an ethics issue said it was simply an overreaction; another called it poor judgment. Universities, more than other institutions, must encourage and protect free speech. Fears of offending alumni contributors or potential public embarrassment may be factors in determining policy, but should not be decisive. No apparent prohibition against particular uses of the bulletin board existed, and instituting a policy of censorship would be a mistake. Any short-term gain achieved would weaken values fundamental to universities and society as a whole.

General Principles

Especially in educational institutions, freedom of speech is critical. It is far more important than avoiding embarrassment or risking negative alumni reaction.

A student should be held to observance of professional ethics as part of his or her education. However, forgiveness should be liberally applied and sanctions limited.

(See also position paper in Appendix.)

SCENARIO IV.9

ATTORNEY: REQUESTING ACCESS TO COMPUTER SERVICE COMPANY RECORDS

John Doe was arrested and charged with rape. The trial was to take place in several months. Mr. Doe had a microcomputer at home and subscribed to a computer service that allowed him to access materials ranging from stock reports and video games to various kinds of literature that can be read on a video display terminal. The prosecuting attorney in Mr. Doe's case sought permission to look at the computer service company's records of Mr. Doe's account. The company monitored subscribers' use of the service (time on and types of materials accessed) in order to bill them. Because the computer service offers a selection of pornographic reading material, the prosecuting attorney hoped to show that Mr. Doe had been spending an inordinate amount of time reading such material. In looking at the account, the attorney would also be able to gather other information about Mr. Doe's personal interests and behavior patterns. He believed all this information would help support his case against Mr. Doe.

Prosecuting Attorney: Seeking Access to Computer Service Company Records of Accused Rapist's Account				
	Total	Unethical	Not Unethical	No Ethics Issue
General Vote	25	4	17	4
Subgroup Vote	7	3	2	2

Opinions

Although the subgroup was split, the majority of the general group considered asking for the service company's records to be legal and not unethical. One group member noted the similarity between reviewing such customer accounts and examining someone's bank records, criminal history, or tax returns. In situations where the evidence cannot be tampered with or destroyed, group members argued for taking the least intrusive approach necessary to achieve the desired result.

Some participants maintained that the request was unethical, even though legal. Obtaining personal information should require a court order; the prosecuting attorney should have to show the probable relevance of the information to the case. Because there is no evidence that pornography has any causal connection to rape, the mere fact of John Doe's accessing pornographic material would not necessarily be relevant.

Some participants who voted that the request was not unethical acknowledged the lack of a demonstrable causal connection between pornography and rape. They pointed out, however, that the company records could well be relevant aside from John Doe's accessing pornographic materials. Transactional records might show the defendant was on line when the rape occurred, and thus he could not have committed the crime. Conversely, the records might show that he had accessed a description of distinctive activities that matched the rapist's modus operandi (e.g., how the defendant "named" his victim).

Service Company: Complying with the Request				
	Total	Unethical	Not Unethical	No Ethics Issue
General Vote	25	17	2	6
Subgroup Vote	7	5	0	2

Opinions

A large majority of group members believed complying with the prosecuting attorney's request was legal, but still unethical. The information maintained by the service company on its clients' use of the services revealed deeply personal aspects of their private lives. That privacy should be respected and protected; the service company should not make the information so easily available, but rather require either the defendant's permission to release the information or a court order. The prosecuting attorney must show that he had reasonable cause to believe that the information would be relevant to the case.

General Principles

Legality and ethics are not necessarily synonymous.

Public servants should seek the least costly, legal way of obtaining essential and material evidence without violating ethical principles.

Companies whose records contain data on individuals' personal habits have a duty to protect that information within the constraints of the legal system.

A computer service company should limit collection of customer information to the minimum necessary for service and accounting. Excessive collection could be unethical.

(See also position paper in Appendix.)

V
Business Practices Including Contracts, Agreements, and Conflict of Interest

Many ethical issues surround contractual agreements for research, consulting services, and software development and enhancements. Ethical conflicts also arise over divided loyalties, such as between country and company. These conflicts raise several important questions:

- Is it acceptable for a researcher to apply money granted for one project toward the completion of another whose funds have run out?

- Should a consultant agree to develop a computer program to a client's specification when it is clearly inferior to a program for the same application developed for another client?

- Is it right to provide consulting services pro bono when other consultants are in business to perform those services for a fee?

- Under what circumstances should a software developer sell the same enhancements to a proprietary operating system to more than one client?

- What is the ethical responsibility of a citizen of one country hired by a multinational or foreign company to provide computer security against intrusions caused by his own country?

In considering these and other questions, some group members were adamant that using funds granted for one project to complete another is unethical unless the funding agency is informed and agrees, even though such practices are commonplace. There are other, acceptable, ways of dealing with insufficient project funds.

Consultants must consider the consequences to themselves of the work they do as well as the consequences to their clients or customers. Ethics cannot (and is not intended to) protect individuals from the consequences of their own business decisions. Neither can consultants ethically withhold a part of the expertise they agree to provide in a contract in order to protect their future business interests. They are obliged to deliver the contracted services.

Professionals should avoid undisclosed conflict of interest situations. They should always endeavor to offer services that are in the best interests of their clients, and should not solicit or accept assignments that they know will produce inferior results. Pro bono work is a well-established, honorable, and perfectly ethical practice. Individuals are within their rights to work for whatever fees they can command, or for nothing if they wish.

Employees are ethically obligated to protect their employers' interests to the extent that those interests are not unethical. One source of conflict is between loyalty to employer and loyalty to country. Where no law prohibits acceptance of contracts to perform work for foreign clients, consultants are ethically bound to observe the contract to which they agreed. In general, the government has an obligation to go through proper channels to retrieve information from private companies, although there may be some situations (i.e., legitimate national security concerns) where the clandestine acquisition of private company information is justified.

SCENARIO V.1[15]

CONSULTANT: PROPOSING AN INFERIOR COMPUTER PROGRAM

Company A invited a consultant to submit a proposal to develop a computer program based on explicit program specifications. The consultant is currently programming the same application for Company B based on far superior specifications that will give it a significant competitive advantage over Company A. The consultant submits a proposal to Company A without mentioning that the specifications are already inferior to the competing product.

Consultant: Submitting Bid on an Inferior Program While Furnishing a Superior Program to Another Client				
	Total	Unethical	Not Unethical	No Ethics Issue
1987				
General Vote	26	18	5	3
Subgroup Vote	8	0	8	0
1977				
General Vote	26	20	5	1
Subgroup Vote	11	10	1	0

Opinions

In 1977, all but six participants agreed that because the consultant had a prior obligation to Company B, and because the companies are competitors in the specified application area, the consultant was unethical and should have refused to submit a bid to Company A, on the basis of a conflict of interest. He should not have proposed an inferior system; that was not in the best interests of his prospective client. If he could not obtain the approval of both

[15]Scenario I.11 in the 1981 book.

customers to provide each of them with the superior product, he should have refused to do business with Company A. The consultant knew he would not be helping his client and would be violating the trust implicitly placed in him.

Some 1977 participants felt that they had insufficient information to judge the ethics of the consultant, although they acknowledged that the consultant was unethical in failing to disclose his relationship with each company to the other. If he had developed the specifications for Company B himself, and if his reputation in the application area was a primary reason for requesting a proposal, then the consultant should not have responded to the proposal from Company A.

One 1977 participant believed the consultant had equal responsibility to both companies; the ethical issues were offsetting. Therefore, he acted inappropriately, but not unethically. Another argued that the situation did not involve any ethical issues. Company A got what it asked for, the consultant was obliged to supply it and did so. Unless information about the efficacy of computer programs of this type was in the public domain, the consultant was not obliged to inform Company A of the obsolete nature of the program it specified.

The 1987 group generally agreed with the 1977 majority that this situation represents an undisclosed conflict of interest. One member cited a case in which a man engaged a lawyer to write an iron-clad will excluding his wife. When he died, the wife hired the same lawyer to break the will. The lawyer clearly had a conflict of interest.

The 1987 subgroup noted that companies using consultants for projects generally have them sign confidentiality agreements, precluding them from disclosing their work for one company to another company. They agreed with the 1977 group's opinion that the consultant's best course, under these circumstances, would have been to decline to submit a proposal on Company A's project.

One dissenter, himself a consultant brought a different perspective to the discussion. Specialists in the computer industry, such as those who work with relational data bases, he contended, will necessarily be doing similar work for several clients. It would be hard to stay in business otherwise. The consultant knows he must be careful not to use proprietary information gained from one client to benefit another client. If a client asks for his best ideas, he provides them without compromising other clients. Yet, if the client gives him specifications, he designs the product to meet those specs on the assumption that the client knows what is needed and why it is needed. As a consultant, he does not second guess his clients, particularly if they do not ask his advice. They may have perfectly sound strategic reasons for developing a product in a certain way that they do not wish to divulge even to him.

Despite this argument, the majority of the 1987 group held to their belief that professional ethics requires that professionals exert their best efforts in their clients' behalf and decline the job if the circumstances do not permit that.

General Principles

Consultants have ethical obligations to offer services that are in the best interests of their clients.

Soliciting or accepting assignments known to produce inferior results or to not accomplish the client's purposes is unethical.

Performing similar work for two competing clients is a conflict of interest unless it is done with both clients' knowledge and approval. Submitting a bid for work that creates an undisclosed conflict of interest situation is unethical.

SCENARIO V.2

COMPUTER SCIENTIST: DIVERTING RESEARCH FUNDS

A computer scientist was the manager and principal investigator of two related research projects. He worked hard on both projects, but the funds on one were depleted before the final report was written. Believing that each project benefited from the work done on the other, the computer scientist allowed his assistant to charge the time required to complete the first project to the second one, but did not inform his manager of this. Subsequently, the scientist also completed the second project by doing considerable unpaid overtime work. The results of both projects were well received.

Computer Scientist: Diverting Research Money				
	Total	Unethical	Not Unethical	No Ethics Issue
General Vote	26	14	7	5
Subgroup Vote	6	5	0	1

Opinions

The majority found the diversion of research funds to be unethical. That many people do it does not make it right or acceptable. Given the circumstances, the computer scientist should have discussed the problem with his manager and perhaps requested permission from the granting agencies to implement his plan. If he had made a good case for the two projects' interrelatedness, most likely permission would have been granted.

Alternatively, the computer scientist could have applied for more funding. In this case he would have been acting ethically as well as avoiding considerable unpaid overtime work. By taking unilateral action, he put not only himself but also his organization at risk. An audit might have revealed the diversion and exposed him and his organization to legal liability and bad publicity, as in the case of time card manipulation by defense contractors.

Those who considered his action not unethical had a variety of reasons. One reasoned that accounting for time and charges is often arbitrary. The second project would have benefited from the computer scientist's getting up to speed on the first project. The first project, therefore, would bear a disproportionate part of the development costs. Another participant considered this issue more a matter of business regulation than ethics. Besides, no harm was done, and both clients were satisfied with the results. In fact, one client (unknowingly) received the benefit of additional work for no additional cost.

Those who voted that this was not an ethics issue also had divergent views. One said that the diversion was illegal and therefore beyond ethical considerations. Another said it was simply poor management. A third believed it was not an issue because it was standard practice.

General Principles

Contractual agreements should be honored.

The ends do not justify the means.

"Everybody's doing it" is no excuse for unethical behavior.

Good contracting and business practices entail anticipating ethical issues, and obtaining informed consent (when necessary) in advance of performing work.

(See also position paper in Appendix.)

SCENARIO V.3

INFORMATION SECURITY MANAGER: PROTECTING MULTINATIONAL COMPANY DATA FROM U.S. GOVERNMENT ACCESS

A multinational company with headquarters in the United States has divisions in many other countries. The manager of information security, a U.S. citizen residing in the United States, was informed by company management that the U.S. government might wish to obtain copies of computerized information in South Africa being communicated to and from the South African division of the company, where an extensive computer system was installed. The manager of information security went to South Africa and installed extensive cryptographic and computer access controls to protect the information communicated and stored in the computer in South Attica. The multinational company's information was thus kept safe from disclosure, but the U.S. government was denied access to information it considered important relative to its diplomatic and economic relationship with the government of South Africa.

Manager of Information Security: Installing Cryptographic and Computer Access Controls to Prevent Access by U.S. Government				
	Total	Unethical	Not Unethical	No Ethics Issue
General Vote	25	1	23	1
Subgroup Vote	7	1	6	0

Opinions

In the absence of valid national security claims, most participants thought the manager's protection of his company's information was not unethical. Managers are charged with protecting their company's interests unless the organization is doing something illegal. Besides, the government can legally obtain access to company information and compel compliance if necessary. Several group members commented that the government has excessively invoked national security interest over the last few years.

The one dissenter considered national interest as outweighing company privacy, even in a situation not involving national security. He questioned, however, the ethicality and legality of a government's intercepting a company's electronic transmissions without justification. Finally, he asked whether an international company has a loyalty to one country and not to the other(s) in which it operates.

General Principles

In the absence of a legitimately documented right to access — usually in the interests of national security — a company and its employees are justified in doing whatever they deem necessary to safeguard access to private company information.

In general, a government should go through proper channels to retrieve information from private companies. However, some situations may justify clandestine acquisition of private company information.

Employees are ethically obligated to protect their employers' interests to the extent that those interests are not unethical. Ethical maturity comes with experience and attention to balancing conflicts that may arise among people's loyalties to country, religion, family, and employer.

(See also position paper in Appendix.)

SCENARIO V.4

SECURITY CONSULTANT: NOT DISCLOSING FOREIGN COMPANY DATA TO U.S. GOVERNMENT

A computer security consultant and citizen of the United States signed a contract, including a confidentiality and nondisclosure agreement, with a client company in Saudi Arabia to assist the company in protecting its information. As a consequence, U.S. government efforts to obtain information about the company were thwarted. The U.S. government complained to the consultant about his work and asked him to reveal what he did. He refused, claiming a duty to his client. The government claimed he had a higher duty to his country.

Security Consultant: Protecting Saudi Company Data from U.S. Government Access				
	Total	Unethical	Not Unethical	No Ethics Issue
General Vote	27	1	25	1
Subgroup Vote	7	1	6	0

Opinions

Given that U.S. law permits U.S. citizens to work for a Saudi company and that the consultant was hired for the express purpose of providing the company with computer security, all but one of the participants found nothing unethical in the consultant's behavior. He was simply adhering to his contractual obligations, and ethically, the contract should prevail in this case. With no compelling national security issues at stake, the U.S. government should not have asked the consultant to breach his contract by disclosing the measures he had used to secure the system. No one should be required to become an intelligence agent.

The one dissenter believed strongly that national interest takes precedence over company loyalty. Some participants agreed that under circumstances where national security was at stake, it might be proper for the consultant to have broken his nondisclosure agreement. Within the subgroup, if this were the case, the vote would have been reversed.

General Principles

Any government should try to obtain needed information without compromising the ethics of its citizens.

When no law prohibits consultants from accepting a contract to perform work for a foreign client, they are ethically bound to observe that contract. When national security is at stake, the government should legally restrict the work that citizens can perform for a foreign client. Professionals thus must be informed of all relevant facts before they make commitments.

(See also position paper in Appendix.)

SCENARIO V.5[16]

SCIENTISTS, BANKS: DECEIVING THE PUBLIC WITH AN AUTOMATIC SPEECH SYSTEM

A team of artificial intelligence scientists has invented an automatic speech-recognition and voice-response system that can be used economically to serve customers of automated teller machines for most banking services. Great effort is made to convince customers that they are conversing with a person. In fact, in case of system failure, a human substitutes for the voice response and follows the same protocol and rules as the system. No capability is included to let the customer know whether conversation is with the device or with a human.

---

[16]Scenario V.1 in the 1981 book.

The project team has complete freedom to exploit the invention. The team patents the invention and forms a company to manufacture and market the product. It is highly successful and widely used in electronic funds transfer systems before any governmental, consumer, or professional agency has an opportunity to study and act on its implications. The scientists-turned-entrepreneurs assume no responsibility for the social implications, claiming their banking customers, who are the users of the product, have that responsibility.

Scientists: Marketing a Device That Could Deceive People and Not Accepting Responsibility for the Use of Their Product

	Total	Unethical	Not Unethical	No Ethics Issue
1987				
General Vote	24	10	9	5
Subgroup Vote	7	0	4	3
1977				
General Vote	26	17	7	2
Subgroup Vote	13	8	3	2

Opinions

In 1987, the general group's perception of marketing a device that might deceive consumers without accepting any responsibility was more evenly divided than the 1977 group's had been. In 1977, a clear majority of both the general and subgroups viewed the scientists' action as unethical. The 1987 group, however, was far from achieving a consensus, perhaps because the possibility of the product is much closer to realization today. Artificial intelligence and expert systems have made great strides over the last 10 years.

The 1987 group seemed to accept the earlier group's contention that the invention itself is not unethical and that marketing it involves nothing intrinsically unethical. They decried the customer banks' refusal to accept responsibility for the way they used the invention, however.

Banks: Using a Device That Could Deceive Their Customers				
	Total	Unethical	Not Unethical	No Ethics Issue
1987				
General Vote	25	13	8	4
Subgroup Vote	7	1	4	2
1977				
General Vote	8	4	4	0
Subgroup Vote	11	11	0	0

Opinions

While the entire 1977 subgroup agreed that the bank acted unethically, the general group was evenly split between calling the action ethical and unethical. In 1987, although the majority of the general group said the bank acted unethically, the majority of the subgroup said the action was not unethical. Several subgroup members found it difficult to believe that people would be deceived by the voice response system. If no one is deceived, then no harm is done, they concluded. This view is inconsistent with the opinion expressed in several other scenarios that it is the intent, not the result, of the action that makes it ethical or unethical. The 1987 general group agreed with the 1977 group that willful deception of the public is unethical.

General Principles

Producing devices or inventions capable of performing humane, possibly human, functions is not unethical. Certain harmful applications of such devices or inventions may be.

Intentionally misleading or deceiving customers is unethical. People have the right to know when and how they are dealing with a machine.

SCENARIO V.6

SOFTWARE DEVELOPER: ENCROACHING ON A CONSULTANT'S BUSINESS WITH AN EXPERT SYSTEM APPLICATION

A software developer produced an expert system shell and offered to hire a consultant to build an application of the system in his field of expertise. The developer sold the expert system application extensively among the consultant's potential clients. The expert system performed well enough that they did not have need for the consultant. As a result, the consultant lost his client market and had to terminate his business.

Variation: The consultant agreed to build the application, but purposely held back some of his expertise in order that the result of his work for the software developer would not destroy the future need for his consulting services.

Software Developer: Selling Expert System to Potential Clients of the Application Developer				
	Total	Unethical	Not Unethical	No Ethics Issue
General Vote	25	1	20	4
Subgroup Vote	7	0	7	0

Opinions

Practically everyone agreed that the software developer did not act unethically in selling the system to the developer's clients. He had contracted with the consultant to develop the application and presumably had paid for the work. The consultant should have known that the application's users would be potential clients and taken action to protect himself. For example, he could have negotiated a contract with royalties, some form of property rights, or limitations on the application to be developed to certain aspects of his expertise. Some participants questioned whether the consultant did in fact retain any ownership of the application. If he did, then it would be unethical for the developer to sell it without recompensing the consultant.

A few viewed the sale as strictly the result of a business arrangement and therefore not an ethics issue. The consultant exercised poor judgment in agreeing to develop the application: Ethics cannot save individuals from the consequences of their own unwise business decisions.

Consultant: Holding Back Some of His Expertise				
	Total	Unethical	Not Unethical	No Ethics Issue
General Vote	25	1	24	0
Subgroup Vote	7	0	7	0

Opinions

Only one member of the group thought holding back expertise would be unethical if the developer knew about it. The dissenter contended that by accepting the assignment, the consultant should do the best job possible. If, on the other hand, the consultant withheld expertise without the developer's knowledge and agreement, 24 of the 25 group members considered that action unethical.

General Principles

Ethics cannot and are not intended to protect individuals from the consequences of their own business decisions. Rather, they are meant to be a standard for the behavior of professionals.

Ethics requires that professionals deliver the product agreed to in the contract.

Secretly withholding part of one's expertise that one has contracted to provide is unethical.

SCENARIO V.7

PRESIDENT SOFTWARE DEVELOPMENT COMPANY: MARKETING A SOFTWARE PRODUCT KNOWN TO HAVE BUGS

A software development company has just produced a new software package that incorporates the new tax laws and figures taxes for both individuals and small businesses. The president of the company knows that the program probably has a number of bugs. He also believes that the fist firm to put this kind of software on the market is likely to capture the largest market share. The company widely advertises the program. When the company actually ships a disk, it includes a disclaimer of responsibility for errors resulting from use of the program. The company expects it will receive a certain number of complaints, queries, and suggestions for modification. The company plans to use these to make changes and eventually issue updated, improved, and debugged versions. The president argues that this is general industry policy and that anyone who buys version 1.0 of a program knows this and will take proper precautions. Because of bugs, a number of users filed incorrect tax returns and were penalized by the IRS.

President: Marketing Software That Has Bugs				
	Total	Unethical	Not Unethical	No Ethics Issue
General Vote	25	24	0	1
Subgroup Vote	7	6	0	1

Opinions

Nearly everyone contended that marketing a software product known to have bugs that may harm users is unethical. They reasoned that when packaged (as opposed to custom) software is sold to the consumer, the developer should have exercised due diligence to find and correct as many bugs as reasonably possible. Any remaining ones should be of the sort that surface only after widespread use. To comply with minimum industry standards, the president should, therefore, have field-tested the product and corrected any deficiencies that were found before offering it for sale. Participants assumed that no field test was conducted and therefore branded his action unethical. It would also have been unethical, having completed testing and finding errors, to market the software without correcting those errors.

The individual who said marketing the product with bugs was not an ethics issue commented that only money was involved.

President: Marketing Product with Disclaimer of Responsibility				
	Total	Unethical	Not Unethical	No Ethics Issue
General Vote	25	19	5	1
Subgroup Vote	7	6	1	0

Opinions

The group's consensus was that although a disclaimer is standard industry practice because of the current legal interpretation of product liability, it is meant to protect developers against unknown product deficiencies, not problems resulting from known bugs. Using the disclaimer in this way is an abuse and therefore unethical.

The minority that voted not unethical stressed the disclaimer's use as standard industry practice.

President: Arguing That His Action Is General Industry Policy			
	Total	Unethical	Not Unethical	No Ethics Issue
General Vote	24	18	2	4
Subgroup Vote	7	7	0	0

Opinions

All members of the subgroup agreed that the president's argument was unethical. Although the majority of the general group concurred, several either considered it not unethical or not an ethics issue. The majority argued that being industry policy does not necessarily make an action right. Moreover, at least one participant suggested that the president in calling his action general industry policy was, in fact, lying. Marketing software known to contain bugs is not general industry policy.

Several participants said that arguing, in and of itself, cannot be unethical unless the arguments used are lies. Freedom of speech guarantees certain rights. They did concede that the president's argument did not contain any valid justification for his actions.

General Principles

Any intentional misrepresentation of a product is unethical.

Professionals are responsible for the foreseeable consequences of the public's use of their products.

An act that is considered to be standard industry practice is not necessarily ethical.

SCENARIO V.8

MARKING MANAGER: DEVELOPING QUESTIONABLE TV COMMERCIALS

The marketing manager of a computer manufacturer developed a series of television commercials claiming, both in the story line and in the voice-over, that a computer will help a company solve problems of poor management. People within the industry, however, debate whether installing a computer will make the situation better or worse. Although the marketing manager was in a position to know this, many people at whom the ads were targeted (first-time users) were not. Some people who bought the computers suffered increased management problems as a result.

Marketing Manager: Representing Computer as a Solution to Poor Management				
	Total	Unethical	Not Unethical	No Ethics Issue
General Vote	25	9	13	3
Subgroup Vote	7	2	5	0

Opinions

Most group members did not believe that making commercials with exaggerated claims was unethical. While granting that such advertising is misleading, they maintained that use of sales hype in advertising is common. One participant commented that this commercial is no worse than 99% of American advertising. Computer manufacturers should not be held to higher standards than other business people. Buyers should discount much of the puffery and not make a decision based on one ad. Those who do not understand computers should seek assistance when purchasing one.

Those participants who considered the ad unethical said that it was misleading, dishonest, and harmful to some purchasers. Being common practice does not make it right. As one pointed out, such false advertising harms not only the consumer but also other computer manufacturers. He reasoned that the vendor promising the most gets the sale. If that product does not deliver as promised, all vendors, even those that practice truth in advertising, are compromised.

General Principles

Truth in advertising is an ethical requirement. That requirement, however, does not preclude imagery, metaphors, and linguistic usage that most potential buyers know how to interpret.

An action that is otherwise unethical is not made ethical by becoming common practice.

(See also position paper in Appendix.)

SCENARIO V.9

CONSULTANT: PERFORMING PRO BONO PROFESSIONAL SERVICES

An independent computer consultant performed extensive planning, analysis, and implementation of computer systems for his clients in the electronics industry at generally accepted rates of compensation. A community hospital and an educational institution both asked him to engage in similar work for them on a voluntary basis. He agreed to do it pro bono, considering the work to be a community service. Other computer consultants who specialized in hospital and educational systems complained to him and his pro bono clients that he was performing without compensation the work they would otherwise have been paid to do. The two clients indicated that they would have been unable to retain those consultants unless the work was performed free of charge. The consultant doing the work concluded that he owed his community a small amount of his time for pro bono work.

Consultant: Providing Professional Service Without Compensation				
	Total	Unethical	Not Unethical	No Ethics Issue
General Vote	27	0	21	6
Subgroup Vote	5	0	4	1

Opinions

No one suggested the consultant acted unethically in providing pro bono services. As one group member asserted, if he is unethical, then so is every philanthropist, volunteer, or person who fixes his neighbor's sink. In fact, said another, professionals should donate some of their services to charitable or community causes. Basically, the consultant has a right to charge or not to charge for his services as he chooses.

Other Consultants: Complaining About Loss of Work

	Total	Unethical	Not Unethical	No Ethics Issue
General Vote	27	15	6	6
Subgroup Vote	5	5	0	0

Opinions

The entire subgroup and the majority of the general group decided the other consultants' complaints were unethical and groundless. One participant cited their approaching the consultant's pro bono clients as being inappropriate. Another said their actions were greedy and mean spirited. If, as they said, the institutional clients did not have the funds to pay for the services performed, no work was taken from the other consultants.

Those who did not see any ethics issue involved nonetheless judged the consultants' actions to be wrong, poor business judgment, and even stupid. In the end, the complainers simply made themselves look bad.

Some suggested that they did not act unethically; complaining, short of harassment, is not immoral or unethical. People have a right to complain when they feel they have been injured, even if their complaints are groundless.

General Principles

Pro bono work is a well established, honorable, and perfectly ethical practice.

It is not unethical to complain, even if the complaint is not justified.

Professionals are within their rights to work for whatever fee they can command, or for nothing if they so choose.

(See also position paper in Appendix.)

SCENARIO V.10[17]

ENGINEER: GAMBLING WITH A CONCEALED COMPUTER

An electronic engineer designed and built a portable, hidden computer that he strapped to his stomach. Input came from switches on the bottoms and tops of his two big toes, and output appeared in an array of colored miniature lights (LEDS) on the inside of the frame of his glasses. The whole apparatus was hidden from view.

He used this computer device to assist him in playing blackjack at gambling casinos in Nevada. Using Thorpe's counting techniques, he had, through the device, an advantage of about 2% over the house. More importantly, he was able to play and use the method without detection. Most card counters can be detected because their mental concentration precludes engaging in conversation. Nevada state law does not prohibit what he did; it only limits use of devices that physically change the odds in a game. However, a casino has the right to expel anyone from the premises without cause.

If the engineer had been caught using his device, he probably would have been told to leave and would never have been allowed to play again. He was never detected and made large amounts of money with the aid of the device. He also made copies of the device and sold them to other gamblers.

---

[17]Scenario I.5 in the 1981 book.

Engineer: Using a Hidden Comptuer to Gamble

	Total	Unethical	Not Unethical	No Ethics Issue
1987				
General Vote	26	9	10	7
Subgroup Vote	10	8	1	1
1977				
General Vote	24	4	15	5
Subgroup Vote	6	0	3	3

Opinions

The 1977 group members believed the engineer violated the equality of opportunity that applies in gaming situations by using an advantage that went beyond mere skill or luck. Though a gun beats four aces, its use is immoral and illegal. Regardless of the ethics of casino gambling, a computer technique for tilting the odds is not in principle different from using marked cards or bribing the dealer. Casino games presume a human player with human limitations. Participants maintained that using deception to gain an unfair advantage is unethical.

The 1977 subgroup was divided on whether use and sale of the device were ethical and legitimate. Those who believed no ethics issue was involved claimed that using and selling the device differed very little from selling the many books on how to win at gambling. The use or sale could involve an ethics issue only if it constituted a fraud or if the casinos explicitly forbade such devices; otherwise, the engineer was simply using publicly available information to improve his position. Those who regarded use and sale of the device as unethical wondered why the engineer kept it hidden. What if the house dealer also used special glasses connected to a computer?

In defending the engineer's actions, some participants argued that blackjack is, comparatively speaking, a game of one intellect against another. Use of the computer did not change this fact, and they did not believe anyone was damaged by the act. The act was therefore not unethical. However, some questioned whether the device gave the gambler an unfair advantage over the house. Players who improve their skills by reading books or learning from others are not regarded as acting unethically. Neither are casinos that sometimes expel individuals only because they have an unusual winning streak. The issue here is equality of opportunity. Is it permissible for a person to give himself an advantage that other people do not have?

The 1987 subgroup took a different perspective; several considered the computer device as an automated aid for carrying out the Thorpe card counting technique. Because the Thorpe technique is an acknowledged ethical and intellectual pursuit that can be carried out secretly, the use of the device is likewise not unethical. Others viewed gambling in Nevada as a business, not gambling, and the engineer as an entrepreneur. Another participant concluded that casinos encourage this type of behavior and deserve to be outwitted.

The wide diversity of opinion found among the 1987 participants seemed to mirror their differing reactions to motorists' use of radar detectors to avoid speed traps. The increase in ethical controversy on this issue since 1977 seems attributable to the increasing technological capability to develop even more powerful devices that have greater miniaturization and are thus easier to conceal.

General Principles

Stealing, even in reprisal, is not ethical and cannot be condoned.

Parties to a contract cannot surreptitiously change the contract. On the other hand, failure to disclose a hidden advantage is not necessarily wrong. Concealment is unethical only when the party has a moral or legal obligation to disclose or not to take an advantage.

The breaking of a law is not ipso facto unethical, particularly when the law itself has no moral basis. Unfortunately, ethics and laws are sometimes unrelated and in conflict. Because gamblers are expected to use their mental processes in one way to beat the house, using them in another way to develop a device for the same purpose is not necessarily unethical.

The use of a hidden device in an accepted relationship is unethical.

VI
Employer/Employee Relations

The relationship between employer and employee represents a special kind of legal contract that does not always have definitive conventions, practices, and trusts. Employees sometimes question the ethicality of their employers' practices and policies — not usually in formal ways, but often with subtle, informal techniques. Similarly, employees are scrutinized by their peers as well as by line managers. Where communications are open and relationships good, conflicts and misunderstandings can often be resolved in a congenial, collegial manner. Where tensions prevail, however, the process of conflict resolution may be much more difficult.

As technology advances, adhering to practices that were conventional just a few decades ago has become more difficult. The rights of an employer to open mail addressed to employees at their place of work, to search employee briefcases and handbags (even in nonsensitive areas), and to monitor employee telephone calls without their knowledge were accepted just a decade or two ago. Today, although these practices still exist and must exist in certain environments, employers may be characterized as being overly autocratic if they persist with such practices.

In the workplace, both employer and employee have rights, but these rights have associated responsibilities and authority. The scenarios in this chapter examine various employer/employee actions and participants' perceptions of and reactions to those actions.

The complexity of the employer and employee relationship raised many questions:

- Is it ethical for an employee to engage in independent work similar to that performed for an employer?

- Should an employee take an internal organizational problem to persons outside the organization? Can an employee be expected to risk personal injury (ostracism, loss of job) by so doing?

- What should an employee do when he disagrees with a company policy? When ethical codes conflict, which takes precedence?

- Under what circumstances is a manager unethical to fire or otherwise discipline employees who violate company policy?

- When is personal use of company resources ethical?

One of the most difficult principles that emerged from the workshop discussion of these questions related to the employees' responsibility in dealing with a problem that their managers refuse to face. When authorities within a company refuse to acknowledge or address a problem, and that refusal has potentially serious consequences, employees may be obliged to take whatever measures are necessary to ensure that the problem is addressed.

Managers are stewards of the company's resources and therefore have the right to determine who can use those resources and under what circumstances. Management has the right to set company policies in general, and to expect that their employees also observe those policies and act in the company's best interests. To safeguard company interests, a manager has the right to monitor employee activities; however, employees have a right to humane working conditions, fair wages for their position and skill levels, and objective performance assessments. They have a right to know, though not necessarily approve, the conditions of their employment, including the possibility that their E-mail or performance is being electronically monitored. With their manager's permission, employees may ethically make personal use of company resources. Unauthorized use, however, especially in violation of an expressed company policy, may be considered petty larceny or even a felony.

SCENARIO VI.1[18]

DIRECTOR, PRODUCT LEARNING CENTER: MARKETING PRODUCTS

The director of a university computer-based learning center proposed to three researchers working for her in the center the formation of a corporation to develop and sell computer-based curriculum materials. The center facilities would be used at night and on weekends, when they were ordinarily idle, so that services could be purchased at a reasonable rate. One copy of any products developed would be licensed to the university without charge. The new corporation would retain all rights to the products and all income from sales.

The researchers declined the opportunity on the ground that the work would be similar to their university work, and the plan would result in many temptations to act unethically. The director berated the researchers for their laziness and lack of enterprise.

---

[18]Scenario IV.5 in the 1981 book.

	Total	Unethical	Not Unethical	No Ethics Issue
1987				
General Vote	24	3	20	1
Subgroup Vote	6	0	6	0
1977				
General Vote	24	13	9	2
Subgroup Vote	10	4	5	1

Opinions

Most 1977 participants agreed that the primary issue of this scenario is that the work required to produce the material that the laboratory director proposed to sell would be similar to her work for the university. Most asserted that proposing a company to produce and market materials similar to those she worked on at the university was unethical.

Almost everyone in 1977 agreed that the researchers' participation in the proposed company would not be unethical. Their participation, however, would open the possibility of unethical behavior. Some participants considered the director's berating her employees for their lack of enterprise unethical.

Although the 1987 participants recognized the potential dangers, they suggested that the steps outlined in the scenario (i.e., getting the university's approval, giving free copies, and purchasing the computer resources used by the corporation) were adequate to protect against conflicts of interest materializing. The majority also maintained that the academic tradition of openness of ideas was an important value at stake in this scenario. Some members reasoned the university was aware of and agreeable to the proposal. As long as proper safeguards were put in place, agreed to, and observed by both parties, the director would not be acting unethically. A few members of the general group agreed with the 1977 majority opinion that the director was unethical — both in forming the company and involving other employees.

General Principles

With an employer's full knowledge and agreement to the arrangement, and taking proper measures to protect against potential problems and conflicts of interest, an employee can pursue outside business projects without being unethical.

In an academic setting especially, professionals are expected to develop materials that can be used by others (e.g., textbooks, training materials, patents).

SCENARIO VI.2[19]

SYSTEMS ANALYST, AIRLINE MANAGEMENT: DISCOUNTING THE NEED FOR INTERACTIVE HUMAN JUDGMENT

The spokesman for a union of airline maintenance workers charged that the airline had introduced a computer program to perform functions that require interactive human judgment if safety is to be ensured. The program is one that schedules maintenance and that reassigns aircraft when emergencies arise because airplanes unexpectedly become unusable.

The systems analyst, under whose direction the program was written, was aware that not all operational factors had been taken into consideration in the program, but he had been assured by management that the decision rules conformed to all the requirements of IATA (International Air Transport Association). In his opinion the program should have been an interactive one, where a person is involved in some of the final decision making, but the company was not prepared to go to the additional expense of an interactive system.

---

[19]Scenario V.3 in the 1981 book.

When testing his program, he could not devise an example where the existing program produced an action that failed to meet a safety condition. Because he could not document reasons for his doubts, and also in part because he was inclined to be defensive about his own work, when he was asked to testify in an inquiry dealing with the union's complaint, he did not volunteer his opinion on how the system should have been designed.

Systems Analyst: Designing a Computer Program That Neglects Known Operational Factors				
	Total	Unethical	Not Unethical	No Ethics Issue
1987				
General Vote	24	10	14	0
Subgroup Vote	6	0	6	0
1977				
General Vote	13	8	5	0

Opinions

In 1977, the general group majority voted that the systems analyst acted unethically in designing a program that omitted some operational factors, even though his manager directed him to do so and assured him that the program conformed with IATA requirements. In 1987, however, most participants contended that the systems analyst did not behave unethically. They focused on the seriousness of the problem and the certainty of its existence. The issue involves possible danger to human lives and property; the consequences, if the program is deficient, are thus unquestionably severe. However, because the programmer tested the program and could not demonstrate any safety problem, 1987 participants argued that his accepting management's assurances was reasonable.

Systems Analyst: Failing to Volunteer His Opinion When Testifying

	Total	Unethical	Not Unethical	No Ethics Issue
1987				
General Vote	24	11	13	0
Subgroup Vote	6	1	5	0
1977				
General Vote	13	7	5	1
Subgroup Vote	8	5	3	0

Opinion

In evaluating the analyst's failure to testify at the inquiry, again the majority in 1987 decided differently than in 1977. The 1977 participants reasoned that the matter was too important for the analyst to withhold his doubts about the system, even though he could not prove a problem existed. The 1987 minority who agreed said that he should back up his opinion, particularly since he was protected by law from retaliation for public testimony. One of the majority noted that, given his inability to document any problem, he should not make himself a pawn for a union assault on management.

Management: Refusing to Take the Analyst's Advice

	Total	Unethical	Not Unethical	No Ethics Issue
1987				
General Vote	23	7	12	4
Subgroup Vote	6	0	5	0
1977				
General Vote	10	3	6	1
Subgroup Vote	7	2	4	1

Opinions

The majority in both 1977 and 1987 agreed that the airline management's refusal to take the analyst's advice was not unethical. The 1987 subgroup members asserted that management was better able than the analyst to understand the operation of the entire system. If the analyst had been able to demonstrate that the program would produce a wrong decision, then the managers should listen. Otherwise, they likely made the proper decision in ignoring the analyst's advice.

General Principles

Ethics requires action that is commensurate with the severity of the danger or potential danger and the degree of certainty that a problem exists that could bring about the danger.

SCENARIO VI.3

RESEARCH TEAM MEMBER: CIRCUMVENTING FEDERAL REGULATIONS

A supercomputer center research team has been working on an important problem involving computerizing air traffic control. A visiting computer scientist from an Eastern Bloc nation appears to have the requisite theoretical knowledge to be of great assistance; however, federal regulations prohibit his involvement. One of the team members discusses the problem informally with him, however, and gains important insight into potential solutions.

Team Member: Discussing Problem with Eastern Bloc Scientist in Violation of Regulations				
	Total	Unethical	Not Unethical	No Ethics Issue
1987 General Vote				
(Knowing)	25	19	3	3
(Not knowing)	25	15	7	3
Subgroup Vote (Either way)	8	7	1	0

Opinions

A significant majority of both groups agreed that discussing a scientific problem with anyone, knowing that it violated regulations, is unethical. They considered whether the team member could have gained important insights into potential solutions without revealing at least the nature of the problem. Subgroup members decided it would be impossible to reveal the nature of a problem without running the risk of communicating exactly what the regulation was created to prevent. Even if the information gathered is useful to solving the problem, they argued the team member has still acted unethically.

A few participants believed informal discussions that avoid transmitting any secrets are feasible. The majority disagreed, however.

The subgroup did not care whether the team member knew that his discussion would violate regulations, although that knowledge did make a difference to a few members of the general group. However, even if the researcher had not heard of the regulation, the majority still believed his action was unethical. It was his responsibility to know of it. Any doubts could have been resolved by checking with someone before the discussion was initiated. One of the dissenters thought it was not unethical to circumvent a ridiculous law. Another participant commented that perhaps more would be gained, even lives saved, by getting insights from the Eastern Bloc computer scientist.

General Principles

People who have agreed to abide by specified conditions of employment (e.g., regulations governing use of classified or proprietary information) are ethically required to do so, all other things being equal. When in doubt as to whether the conditions apply in certain circumstances, they are obliged to clarify the situation before acting.

SCENARIO VI.4

CONSORTIUM: DENYING NONMEMBERS AVAILABLE TIME ON SUPERCOMPUTER

With federal funds, a consortium of universities has established a networked supercomputer center. The policies of the center allow researchers at these institutions first priority for its use. The center manager evaluates requests for use. He routinely rejects requests from other institutions, without evaluating them for scientific merit, once the 20% time set aside for other institutions is taken, even though available time may remain of the 80% set aside for consortium members.

Center Manager: Rejecting Requests from Nonconsortium Institutions for Additonal Time on the Supercomptuer				
	Total	**Unethical**	**Not Unethical**	**No Ethics Issue**
General Vote	20	6	8	6
Subgroup Vote	7	0	1	6

Opinions

The subgroup nearly unanimously agreed that the manager's routine rejection of requests was not an ethical but a management issue. Those in the general group who said the center manager's action was not unethical reasoned that he would have no right to break the rules laid down by the center's management. Were he to do so, he would be acting unethically. His one reasonable recourse, they contended, would be to have the rules changed to allow more flexibility in allotting computer time. For economic or other reasons, management may not wish to allocate more time for nonconsortium use, even if the computer were standing idle. For example, management may want to ensure that the price remains high. In any case, the center manager was right to follow management policy in operating the center.

One of the participants who thought rejecting nonconsortium requests beyond the 20% was unethical said that blindly following such rules should be actively discouraged. People should use common sense in applying rules.

General Principles

Management has a right to set policy for the company and to expect staff to obey that company policy unless doing so would be unethical or illegal.

Employees implicitly, if not explicitly, agree to observe company policy as a condition of their employment.

If employees disagree with a company policy, they should try to have it changed; in the meantime, however, they should continue to observe the policy.

SCENARIO VI.5

RESEARCH CHEMIST: USING BULLETIN BOARD SYSTEM (BBS) FOR RECRUITING

A group of research chemists in the plastics industry, all working on similar problems, decided to develop an electronic bulletin board system to exchange technical information on a timely basis. The BBS was run by a chemist in one company and made available to all the others. This chemist was approached by a recruiter in his company who indicated a need to hire research chemists in his field. The chemist and recruiter developed employment promotional materials and used the BBS to disseminate them to chemists working at the other companies. All the companies had a policy prohibiting use of the U.S. mail to recruit chemists directly at their companies. Because the BBS did not qualify as a postal system, it was not covered by the policy. A chemist was hired from another company as a result of the recruitment effort. The other company complained that this was unfair.

Recruiter: Asking Chemist to Use BBS to Help Recruit Needed Chemists				
	Total	**Unethical**	**Not Unethical**	**No Ethics Issue**
General Vote	26	15	8	3
Subgroup Vote	7	7	0	0

Opinions

The majority of the general group and the entire subgroup agreed the recruiter was acting unethically. Although the recruiter was not violating the letter of company policy, he was certainly violating its intent or spirit. His action ran counter to the concepts underlying restrictions on the use of the postal service (e.g., no raiding). One group member noted the similarity between using the BBS and the postal service: both involve employees being contacted at their workplaces.

In asserting the action was not unethical, one participant found nothing wrong in the recruiter asking the chemist, although the chemist should not have used the BBS; the recruiter was just doing his job. Others considered the rule to be overly specific; it focused only on the postal service and no other communications media. The managers of the company were at fault for not being more comprehensive in defining policy (see below.)

Chemist: Using the BBS for Purposes for Which It Was Not Intended				
	Total	Unethical	Not Unethical	No Ethics Issue
General Vote	25	18	7	0
Subgroup Vote	7	7	0	0

Opinions

The subgroup members unanimously found the chemist's use of the BBS for recruiting to be unethical for the same reasons they cited for the recruiter. In the general group, some who had excused the recruiter's action by saying that he was only doing his job agreed with the majority in the case of the chemist. The chemist's use of the BBS was not only counter to the companies' recruiting policy, but counter to the purposes for which the BBS itself was established. Several group members commented that he should have asked for and received the consent of the other participants before agreeing to disseminate recruiting materials over the BBS.

The minority, who called the action not unethical, continued to maintain that the company did not specifically prohibit use of the BBS and therefore its use was proper. One asked rhetorically if anyone knew of an ethical maxim stating that things should not be used for purposes other than those for which they were intended.

Company Managers: Setting Narrowly Defined Policy				
	Total	Unethical	Not Unethical	No Ethics Issue
General Vote	25	6	2	17
Subgroup Vote	7	5	0	2

Opinions

On this issue, the majority of the general group said that narrowly defining policy was not an ethics issue, while the majority of the subgroup called the action unethical. The majority seemed to agree, however, that the restrictions on recruiting amounted to unfair restraint of trade and inhibited employees' freedom of movement. Almost everyone agreed that the company managers were at least negligent in having a policy that could be so narrowly interpreted.

General Principles

Company policy should state general principles rather than narrow, specific examples of actions that are not permissible.

Professionals should recognize and observe the intent as well as the letter of legitimate company policies.

An employee need not apply specific, restrictive, perhaps illegal, rules to a broad spectrum of other activities.

(See also position paper in Appendix.)

SCENARIO VI.6

INFORMATION SECURITY MANAGER: MONITORING ELECTRONIC MAIL

The information security manager in a large company was also the access control administrator of a large electronic mail system operated for company business among its employees. The security manager routinely monitored the contents of electronic correspondence among employees. He discovered that a number of employees were using the system for personal purposes; the correspondence included love letters, disagreements between married partners, plans for homosexual liaisons, and a football betting pool. The security manager routinely informed the human resources department director and the corporate security officer about these communications and gave them printed listings of them. In some cases, managers punished employees on the basis of the content of the electronic mail messages. Employees objected to the monitoring of their electronic mail, claiming they had the same right of privacy as they had using the company's telephone system or internal paper interoffice mail system.

Information Security Manager: Monitoring Electronic Correspondence of Employees				
	Total	Unethical	Not Unethical	No Ethics Issue
General Vote	25	14	11	0
Subgroup Vote	5	1	3	1

Opinions

Monitoring E-mail was unethical according to a narrow majority of group members because the employees apparently had no warning that they might be monitored or that their use of the system was restricted. They were thus less likely to send discreet messages. Under the circumstances, management was taking unfair advantage of the employees. Furthermore, some participants argued that E-mail should be given the same rights to privacy as postal mail.

The sizeable minority who voted that the information security manager did not act unethically noted the considerable differences between U. S. postal mail and E-mail. E-mail, as well as telephone systems and internal mail in the workplace, are company property and resources. No rights of privacy extend to individuals using property or resources supplied for company (as opposed to personal) business. E-mail is also by its nature more accessible than other forms of communication. Finally, they argued, monitoring all communication forms is necessary for management control of its resources, particularly when a company has trade secrets or classified information to protect.

Information Security Manager: Informing Management of Abuse				
	Total	Unethical	Not Unethical	No Ethics Issue
General Vote	14	22	2	0
Subgroup Vote	6	6	0	0

Opinions

Giving management printouts of the actual messages is quite different. Although a summary of the types of calls made would be acceptable, all of the subgroup and all but two of the general group considered it highly unethical to provide others with the detailed content of clearly personal messages. The right to privacy is the central ethical principle. Material security considerations do override privacy considerations, but what the security manager is reporting has nothing to do with security (unless the relationships make individuals vulnerable to blackmail). Even then, in the absence of a prohibition on the personal use of the E-mail system, providing the messages verbatim to management is unethical.

The two individuals who did not consider the action unethical reiterated that employees have no right to expect privacy in E-mail. The facility is meant for business communications and is subject to management scrutiny.

Employees: Using Electronic Mail System for Personal Communications				
	Total	**Unethical**	**Not Unethical**	**No Ethics Issue**
General Vote	25	10	11	4
Subgroup Vote	6	1	3	2

Opinions

The general group was divided on the ethicality of using E-mail for personal purposes. More than half the group believed that type of use was not unethical or not an ethics issue based on the absence of a clearly stated company policy forbidding the use of E-mail for other than company business. Several participants asserted that if the company explicitly prohibited personal use, the employees would be acting unethically. Such use would also be unethical if it interfered with the employees' job performance. Otherwise, according to some, personal use of E-mail might be indiscreet or unwise, but not unethical. Employees should realize that the system can be monitored and, if security considerations are involved, probably is.

Some participants characterized betting information as not personal or private, particularly given its illegality.

A sizeable minority of the general group contended that employees' personal use of corporate resources was unethical prima facie. It is a misuse of corporate facilities and could reduce system effectiveness. One participant believed this use could constitute petty larceny or even a felony. Unless a clear policy permits personal use, employees should treat E-mail resources the same as petty cash or a company car.

Information Security Manager: Failing to Ask for Rules on Personal E-mail Use from Management

	Total	Unethical	Not Unethical	No Ethics Issue
General Vote	24	11	3	10
Subgroup Vote	6	0	6	0

Top Management: Failing to Set Rules and Inform Employees

	Total	Unethical	Not Unethical	No Ethics Issue
General Vote	23	20	2	1
Subgroup Vote	6	5	0	1

Opinions

The majority opinions on rule setting reflect the relative obligations and responsibilities of the information security manager and top management. These participants argued that management is responsible for setting clear policies, defining the consequences of violating them, and ensuring that each employee understands them. Depending on a policy's importance, management may even insist that employees sign an acknowledgment that they have been briefed and understand the policy. The policy should then be applied consistently. Therefore, most group members believed enforcing an unstated or fuzzy policy that violates employees' apparent expectations of privacy is unethical. Those who disagreed iterated that E-mail by its nature is not private and cannot be compared to the U.S. Postal Service.

Conversely, a majority argued the information security manager is not responsible for ensuring that policies are set or promulgated, but only for carrying them out. For this reason, all the subgroup maintained that he did not act unethically, and a majority of the general group decided that his failure to ask for a rule on personal E-mail use was either not unethical or no ethical issue was involved. A significant minority did think the information security manager was ethically obliged to ask for a clear, publicized policy from management. In its absence, however, he is acting unethically by monitoring the E-mail because the lack of a policy infers that the employees' privacy will be respected.

Top Management: Punishing Some Employees Based on the Content of Their E-mail Messages				
	Total	**Unethical**	**Not Unethical**	**No Ethics Issue**
General Vote	25	23	0	2

Opinions

Almost everyone contended that imposing a penalty based on message content was unethical, for several reasons. First, employees apparently did not know that their E-mail would be monitored. It would be appropriate to punish employees for violations of a clearly stated, and well-promulgated company policy, but not for one they were unaware existed.

Second, because management knew the actual contents of the messages, they seemed to be applying employee sanctions based on behavior indicated by the message content rather than for misuse of E-mail. Management acted improperly in disciplining employees for non-job-related aspects of their personal lives. This was a violation of the employees' right to privacy and freedom of action in their personal lives.

Third, because employees were punished "in some cases," management again appeared to be acting on the content of the message rather than the unauthorized use of company resources. Such disciplinary actions are also arbitrary. Management should be consistent in applying sanctions.

If the information security manager had given management a simple list of those who were using E-mail for personal reasons, if the employees knew such use was against company policy, and if penalties were applied consistently, then according to the participants, management would be acting well within its authority. Otherwise, management acted unethically.

General Principles

Employers have a right to limit or prohibit employees' use of company resources. However, these limits must be clear, must be made known to all employees in advance, and must be consistently enforced.

In the absence of a policy forbidding use of E-mail or other intraoffice communications, reading communications intended for others is a violation of privacy.

It is wrong to monitor employees' personal communications without their knowing that monitoring is at least a possibility. It is an invasion of privacy. However, employees' consent, beyond their deciding to continue their employment with the company, is not necessary.

When using company E-mail, individuals should realize that by nature it is not private. It is inherently more accessible, more open to others' perusal than other intraoffice mail.

Employers have no right to impose sanctions on employees for non-job-related behavior.

When personal habits interfere with job performance, they become legitimate management concerns. Otherwise, employees have the right to privacy and freedom of action in their personal lives.

A company has the right to engage in surveillance of employees when suspicions of explicit wrongdoings cannot prudently be ignored.

Unauthorized use of company resources may be construed as petty larceny or even a felony, e.g., California Penal Code Section 502 — unauthorized use by an employee of over $100 of computer services.

A security officer has a right and duty to act on information clearly inimical to his employer's legitimate and ethical business interests.

(See also position paper in Appendix.)

SCENARIO VI.7

EMPLOYER: MONITORING AN INFORMATION WORKER'S COMPUTER USAGE

An information worker in a large company performed, her assignments on a workstation connected to the company's mainframe system. The company had a policy of allowing employees to use the computer services for personal purposes as long as they had the explicit approval of management. The woman had such approval to use the system for the extracurricular recreational activities of the employees in her department.

The company suspected a rising amount of employee unrest because of its potential acquisition by another company. Management had the security department monitor all computer service activities of the information worker. Memos, letters, E-mail messages, bulletin board notices, collections and expenditures of money, and budgets were all carefully scrutinized for evidence of employee unrest. In addition, the security department prepared reports detailing the information worker's use of the computer services — both her regular work and her employee recreation work. These reports were read and analyzed by a wide range of company managers and were stored indefinitely in company vital records facilities. All of this took place unknown to the information worker.

Top Manager: Allowing Employees to Use Computer Services for Approved Personal Purposes

	Total	Unethical	Not Unethical	No Ethics Issue
General Vote	28	0	9	19
Subgroup Vote	5	0	0	5

Opinions

The subgroup unanimously agreed with the majority of the general group that no ethics issue was involved in the decision to allow or prohibit the use of office computers for personal purposes. Because the computer services belonged to the company, its managers could determine how the equipment would be used. By approving of such personal uses, they are in effect providing another company benefit to the employees. Those who described the policy as not unethical essentially agreed with the majority. No one considered this action to be unethical.

Top Manager: Directing Security Department to Monitor Computer Services Activities				
	Total	Unethical	Not Unethical	No Ethics Issue
General Vote	27	24	1	2
Subgroup Vote	5	3	1	1

Opinions

A substantial majority believed monitoring the computer services used without the employee's knowledge was unethical. Although employers argue for controlling everything in the workplace because they pay for it, this particular action is an unjustifiable violation of employee privacy.

At issue is the principle of respect, participants insisted. The employer/employee relationship is essentially a contract. For a contract to be ethical, both parties to the agreement must understand what they are agreeing to and freely choose to make the agreement. Furthermore, because management controls the workplace (i.e., makes the rules/policy), management is in a better position to explain the conditions of employment than the employee is to ask all the right questions. If management informed the workers that they could, with permission, use the computer services for personal activities, and at the same time advised them that these uses might be monitored, then the action would not be unethical.

Those who considered it not unethical to monitor the employees secretly said that the action was justified by security concerns over the merger. Those who decided that no ethics issue was involved based their decision on management's prerogatives within the workplace.

General Principles

Employers have a right to set policy. However, when the policy directly affects employees, the employer must make them aware of the relevant terms of their employment. To do otherwise is unethical.

Workplace surveillance without employees' knowledge to detect "unrest" is an invasion of privacy and highly unethical.

(See also position paper in Appendix.)

SCENARIO VI.8

COMPANY MANAGER: FORBIDDING COMPUTER GAMES

The manager of research in a computer company explicitly stated that the company's computers were not to be used for playing games and said that anyone found in possession of games software would be subject to dismissal. On a random inspection of files, the manager found a game in a programmer's file, and the employee was punished.

Manager: Prohibiting Computer Games in Employees' Files			
	Total	Unethical	Not Unethical	No Ethics Issue
General Vote	23	1	3	19
Subgroup Vote	8	0	0	8

Opinions

Only one individual objected to a manager forbidding computer games, specifically to the manager's making a unilateral decision.

The other participants agreed, whether they categorized the action as not unethical or not an ethics issue, that controlling use of company resources was a management prerogative. The computer was acquired for business not personal use.

Manager: Stating That Possession of Games Is Sufficient to Warrant Dismissal				
	Total	Unethical	Not Unethical	No Ethics Issue
General Vote	25	7	5	13
Subgroup Vote	8	0	3	5

Opinions

As before, the majority argued that the penalty for game possession, although harsh, was a business decision. As long as company policy is clear, that is, all employees are aware of it, no ethical issue is involved. Those who decided the penalty was not unethical based their assessment on the same reasoning.

A vocal minority disagreed. One participant insisted the punishment must fit the crime or it is not ethical. Firing, he said, was far too severe for simply having a game file on the computer. Several others concurred. Some participants also noted that simple possession was not necessarily proof that the individual had used the computer to play the game.

Manager: Inspecting Employee Files for Games

	Total	Unethical	Not Unethical	No Ethics Issue
With Employee Knowledge				
General Vote	24	6	17	1
Subgroup Vote	8	1	7	0
Without Employee Knowledge				
General Vote	24	20	4	0
Subgroup Vote	8	7	1	0

Opinions

If the employees are notified that files will be inspected, the majority agreed that inspecting them would not be unethical. One participant noted the analogy between employees' electronic files and paper files; the latter are not ordinarily subject to random inspection and can contain personal material. In general, however, electronic files are more open, more easily accessed by others. Often, it is not as easy to secure them as it is to lock a file cabinet or desk drawer. Furthermore, no norms have as yet been developed for electronic files.

Those who considered random searches unethical even with the employees' knowledge insisted that such actions are an invasion of privacy — electronic files should be regarded in the same way as conventional files.

If the employees were not notified of the policy, participants voted differently. Most group members would then consider random surveillance to be improper and unethical. One dissenter pointed out that in making the rule and setting the penalty, the manager in effect served notice that files would be searched. There would be no other way to ascertain whether the rule was being observed.

Manager: Punishing Employee for Possession of Game				
	Total	Unethical	Not Unethical	No Ethics Issue
General Vote	25	8	12	5
Subgroup Vote	8	2	4	2

Opinions

About half of both groups considered the manager not unethical in punishing the employee. Another fraction maintained that the punishment was not an ethics issue.

Several people questioned whether possession of the game should carry with it a presumption of guilt in the same way that possession of burglar's tools would. If the manager's policy is ethical, however, then carrying it out would not be unethical. In fact, not following through would be poor business practice. Rules that are not enforced will not be taken seriously.

General Principles

Management has the right to control and set policies for employees in their use of company resources.

Management also has the right to determine the consequences of policy violations and to enforce those consequences, as long as both the policy and the consequences follow the principle of due process and are clearly stated to the employees.

Management's randomly searching employee files, whether electronic or conventional, is unethical if the policy has not been clearly communicated to employees.

(See also position paper in Appendix.)

SCENARIO VI.9

AIRLINE EXECUTIVE: AUTOMATICALLY MONITORING THE PERFORMANCE OF AIRLINE TICKET AGENTS

An airline executive, who was in charge of all personnel issuing tickets at the airline's terminal counters, had a new computer program developed for the centralized reservation system to monitor the work and performance of the sales and ticket agents. The program provided detailed performance statistics on each agent, including the number of sales and how quickly and accurately an agent produced airline tickets. It also monitored and reported on time spent working at agent ticket counters. This monitoring was not revealed to the sales and ticket agents, but the statistics were used extensively in job performance evaluations for salary adjustments and advancement. (In a second version of this scenario, the monitoring was revealed with the same results.) The result was that far fewer supervisors were needed, giving the sales and ticket agents much more freedom from human supervisory monitoring and more objectively applied rewards for high performance.

As the sales and ticket agents realized that their work and work habits were in some way being monitored, however, they became stressed, felt alienated from management, and devoted less time and were less careful and friendly in serving airline passengers. In job performance reviews, special circumstances or personal needs of the sales and ticket agents tended to be ignored in favor of purely statistical measurements as the criteria for job advancement.

Airline Executive: Using Computer Program to Monitor Ticket Agent Activities				
	Total	Unethical	Not Unethical	No Ethics Issue
Without Employee Knowledge				
General Vote	27	21	5	1
Subgroup Vote	5	2	3	0
With Employee Knowledge				
General Vote	27	4	18	5
Subgroup Vote	5	1	2	2

Opinions

The employees' knowledge of their being monitored affected whether participants believed the action was unethical or not. Group members reasoned that monitoring workers' performance is acceptable; all production jobs are measured in some way. However, employees have the right to know that they are being monitored and the basis on which they are being judged. Therefore, if done secretly, the monitoring is unethical. If done openly, the monitoring is not unethical because the ticket agents know what is happening. They then could find other positions, continue their present job and accept the monitoring process, or negotiate the monitoring process during collective bargaining and strike if a satisfactory agreement could not be reached.

Those who believed the monitoring was unethical regardless of whether it was done openly or secretly cited management's responsibility to provide humane working conditions. One participant mentioned the negative effect on the customers as sufficient to make the action unethical.

People who did not view the monitoring as an ethical issue agreed with the majority that employees should expect to be evaluated. Yet, they noted that the managers were foolish to continue a practice that was achieving the opposite of the ultimately desired effect — better customer service. Although individual performance was being statistically measured, the service to the customer was becoming worse, not better.

Airline Executive: Using Program Data for Personnel Decisions to the Exclusion of Other Measures

	Total	Unethical	Not Unethical	No Ethics Issue
Without Employee Knowledge				
General Vote	27	17	2	8
Subgroup Vote	5	1	0	4
With Employee Knowledge				
General Vote	26	13	6	7
Subgroup Vote	5	1	1	3

Opinion

The majority of the general group in both 1977 and 1987 considered the exclusive use of program data for personnel matters unethical whether information is gathered openly or secretly, while the majority of the subgroup viewed this action not as an ethics, but as a management issue. The general group argued that using only this one measure of performance was intrinsically unfair and inadequate. Other relevant factors should be considered in an employee's evaluation. Most of the subgroup thought such use was not an ethics issue. Although the action may represent poor judgment, bad management, or even stupidity, managers can use whatever legal review criteria they believe are appropriate and equitable.

Airline Ticket Agents: Devoting Less Time and Care to Serving Airline Passengers

	Total	Unethical	Not Unethical	No Ethics Issue
Without Employee Knowledge				
General Vote	27	7	3	17
Subgroup Vote	5	2	0	3
With Employee Knowledge				
General Vote	27	7	3	17
Subgroup Vote	5	2	0	3

Opinions

On the issue of diminished service to customers, whether the agents knew they were being monitored did not affect group members' opinions. A clear majority of the general group and a small majority of the subgroup decided that providing worse service was not an ethics issue, but perhaps a normal response to unfair treatment. Whether the ticket agents could be regarded as professionals evoked considerable discussion. If they are not professionals, they cannot be held to professional ethical standards. Less can be expected of them. Professionals, on the other hand, have a greater obligation to adhere to high ethical standards.

A minority of the participants believed the agents were acting unethically. Although the agents' emotional reaction might be understandable, it cannot be justified. Employees, professionals or not, have at least an implied agreement to work for their employer's best interests. The ticket agents should try to redress the situation through proper channels, not by taking out their frustrations on the customer.

General Principles

Employers have the right to monitor their employees for purposes of evaluating their performance.

Professionals should not be excused for giving less than their best efforts to their customers because their managers treat them unfairly.

Employees have a right to know the criteria being used to evaluate their performance.

Using only one source of data to make personnel decisions when other valid measures are available may be expedient but not ethical.

(See also position paper in Appendix.)

SCENARIO VI.10

HUMAN RESOURCES INVESTIGATOR: SEARCHING EMPLOYEE RECORDS TO DETECT DRUG USE

A human resources investigator in a large company was given the assignment of identifying employees suspected of on-the-job drug use that was resulting in a high percentage of product failures. Management expected him to use the conventional investigation methods of observation, interrogation, and use of undercover informants. Instead, and unknown to management, he developed and used a company program to systematically search the personnel and timekeeping records of all the employees whose jobs were related to product quality. Using statistical tabulations and cross-tabulations, he was able to identify those employees most likely to be engaged in drug use according to medical leave records, tardiness, poor performance reviews, and the results of physical examinations employees underwent voluntarily for a one-time survey performed by an outside medical research group. From these data and analyses, he identified the employees suspected of drug use. All the suspected employees were put on notice that after a one-month grace period they would be subject to urinalysis to detect the presence of drugs.

Managers: Telling Human Resources Investigator to Identify Employees Suspected of Drug Use				
	Total	Unethical	Not Unethical	No Ethics Issue
General Vote	27	12	11	4
Subgroup Vote	5	2	1	2

Opinions

Both groups were split on whether managers were ethical in ordering the human resources investigator to identify suspected employees. Those believing the action was unethical asserted that the managers should not be concerned with investigating drug use (the possible cause of product failures), but with job performance (the result). Job performance is company business; drug use is a crime that the police should handle. The managers could achieve the desired result — a lower product failure rate — by taking general measures to reduce tardiness, absenteeism, low productivity, and poor attitude. They should not take on quasi-police functions, however.

Those who considered the managers' action not unethical argued that the company is obliged to provide a safe workplace and a good product. Depending on the product, product failure can be dangerous to the public. At a minimum, product failure adversely affects the company's reputation and profitability, which managers should act to protect. Thus, the managers had the right to determine whether employee drug use was the reason for product failures, and if so, to take action against drug users.

Several participants thought no ethics issue was involved. The managers' action in ordering the investigation was simply normal business practice, they contended.

Investigator: Using a Computer Program to Search Employee Records for Signs of Drug Use

	Total	Unethical	Not Unethical	No Ethics Issue
General Vote	27	14	7	6
Subgroup Vote	5	1	3	1

Opinions

The use of a computer program to search personnel records for signs of drug use elicited clear, but opposite majority opinions in the general group and subgroup. The general group majority contended that the action was unethical because the employees' privacy was invaded, particularly with regard to the use of the voluntary, one-time physical examination data. Using this information for other than the originally stated purpose and without the employees' knowledge or permission was clearly a breach of confidence. Some participants also questioned the accuracy of the investigator's results. Employees might be wrongly accused on the basis of invalid research. One group member also maintained that the investigator's not telling management that he planned to use a different method was unethical.

Those who decided the computer program use was not unethical (a majority of the subgroup), maintained the investigator's methods were perfectly legitimate, even ingenious. The records he used belonged to the company and the managers had the right to use them. One participant noted that he avoided the need to use the more conventional, but deceptive, investigative technique of using undercover informants. Several group members indicated that the employees should have been informed of the investigation. Some accepted the general use of company files but objected to using the medical survey data.

The participants who believed that no ethics issue was involved called the investigator's action an effective use of company information. One member asked rhetorically whether the use of a computer automatically makes an investigation unethical.

Managers: Telling Identified Employees They Would be Subject to Urinalysis				
	Total	Unethical	Not Unethical	No Ethics Issue
General Vote	22	8	9	5
Subgroup Vote	3	1	0	2

Opinions

A number of participants, albeit a minority, decided that forcing certain employees to take urinalysis tests was unethical because the company was exercising quasi-police powers and dispensing with due process. They argued that such an action was "harassment by statistics." Basically, the company considered these individuals to be guilty of drug use, thus damaging their reputations, without real evidence and without giving them an opportunity to defend themselves. Group members also contended that testing employees selectively was discriminatory. A more acceptable policy would be to require all, or randomly selected, employees to be tested. One person suggested that instituting a drug counseling service to address the problem would have been better.

The largest segment of the group believed the managers' action was not unethical, assuming the testing was done competently, because the requirement of probable cause does not necessarily extend to private situations. The one month's notice, in effect an opportunity to stop using drugs if necessary, was fair. The managers could have insisted on immediate testing. Participants also noted that damage to the reputation of innocent individuals could be avoided if the individuals were notified by private letter, and discreet arrangements were made to carry out the testing.

General Principles

Informed consent should be obtained before personal information is used for a purpose other than that for which it was originally collected.

Policies should be fairly applied to all employees.

The use of a computer to replace a manual activity, does not, per se, change the ethical nature of the activity.

Management has the right to investigate the direct and known causes of product failures or diminished performance of employees and to rectify the causes.

(See also position paper in Appendix.)

VII
Summary of Ethical Issues

The 1981 book, Ethical Conflicts in Computer Science and Technology, made some tentative steps toward developing a set of ethical principles for computer professionals based on the computer science and technology ethics workshop convened by SRI International under a grant from the National Science Foundation. The 1987 study further explored and developed principles for ethical conduct in the field. A comprehensive codification of ethical principles will require considerably more study and discussion than was possible within the limits of these two workshops; however, some guidelines for ethical conduct can be derived. To make them more easily accessible, SRI has grouped them into four increasingly narrow categories according to whom the principles most apply: general public, professionals, employers, and employees.

GENERAL PUBLIC

The general principles identified are simple, common truths that apply to everyone's behavior. We all know that the ends do not justify the means, and that two wrongs do not make a right. Most children have tried to justify an action or request with an "everybody's doing it." Such cliches are no excuse for harmful actions.

Responsibility

A key concept that arose during the workshop discussions is that ethics requires intent. If the intent to act or not act is not present, no ethical issue is involved. In ethics, responsibility requires the individual to be knowledgeable of the ethicality of acting or failing to act in a given circumstance.

Ethical behavior is sometimes neither clearcut nor easy, partly because ethical obligations take many forms. Our ethical obligations to ourselves, our families, our employer and profession, customers or clients, church, community, and country may conflict (for example, loyalty to country over loyalty to employer). Determining which takes precedence in a given situation can be difficult. This complexity of ethical issues is one reason (along with the increased possibility of harm) that society places a greater ethical burden on adults, professionals, and those in positions of public trust than it does on children, students, or those with limited education and experience.

Regardless of age, education, or position, however, individuals are expected to defend their values and ethical principles. Society does not always perceive acting on high personal principles as ethical; for example, acts of civil disobedience may be illegal or may violate company or other organizational rules. Yet, violation of a law or rule is not necessarily unethical, particularly when the law itself has no moral basis. Conversely, having the legal right to do something does not mean that the action itself is ethical.

Workshop participants recognized several broad principles concerning the public's ethical responsibilities:

- If wrongdoing or harm is suspected, failing to act is unethical.

- Use of data obtained unethically is also unethical.

- Avoiding designation as the responsible party for a task is not unethical, but evading one's responsibility when it has been assigned, and one has accepted it, is.

Several specific points concerning our responsibility to protect our rights and interests, as well as those of our employer, clients, customers, community, and country, were also made:

- Individuals must accept responsibility for protecting their own legitimate interests; they cannot expect others to watch out for them.

- Although incompetence (inability to recognize a problem) is not unethical, neglecting recognized responsibilities is.

- Finding and reporting a system weakness is not a license to take advantage of it.

- Offering a service that has both risks and benefits is not unethical, as long as the risks are known by those who choose to use the service.

- Violation of a trust (e.g., compromising an employer's computer system to demonstrate inadequate security) can be countenanced only when the act is undertaken in the interests of the community and is promptly communicated to the appropriate authorities. To delay makes the motivation seem specious. Those taking the law into their own hands must be willing to assume the risks and take the consequences.

- To achieve a higher good for society, individuals may need to make a personal sacrifice. Whistle blowers, for example, may not only lose their jobs but also be ostracized by other employers. Unfortunately, society cannot be expected to compensate individuals for damages incurred in its behalf. The ethicality of this societal injustice or how it might be addressed was not explored.

Fees

Individuals have a right to charge whatever the market will bear for their services, or to offer them free if they wish. Pro bono work is a well-established, honorable, and ethical practice. The fact that others charge for performing that service or providing that product, or that in other circumstances the person volunteering services would ask for payment, does not mean that ethically a fee must always be charged.

Contracts

Good contracting, budgeting, and management practices provide a significant degree of safety to individuals having to confront ethical issues. Well-written contracts protect all parties concerned. Contracts, whether they involve rules for games or the exchange of goods and services, can greatly reduce both the opportunities and the motivation for unethical behavior. In particular:

- The rules for bidding on a contract should cover all cases and choices of action and be made known to all parties involved.

- A good contract will help to avoid ethical conflicts.

- Parties to a contract must not surreptitiously change it, although failure to disclose a hidden advantage during negotiations is not necessarily wrong. Concealment is unethical only when a moral or legal obligation to disclose exists.

- Once made, contractual agreements must be honored.

Workshop participants concluded that acting with courtesy, although not a true ethical issue, helps to ensure ethical behavior.

Privacy

The right to privacy is guaranteed by common law and some state laws (e.g., California). Many regard it as a pivotal right needed to ensure a civilized and free society. The importance of respecting and protecting others' rights to privacy was a recurring theme during the workshop. Organizations, as well as individuals, have a right to privacy. As with every right, the discussion also recognized limits to that right. For example, in accepting and using a credit card, individuals relinquish the right to keep their credit history private from those extending them credit. The company extending credit has an overriding right to be assured that those to whom it gives credit have the ability to pay. Nevertheless, the credit company also has a responsibility to exercise care in its use of that information.

In another area, as several political campaigns have demonstrated, individuals who run for office must expect to have their backgrounds scrutinized more closely than private citizens. The public's right to know the character of those they elect takes precedence.

Several more specific principles concerning rights to privacy also emerged during the workshop:

- When the confidentiality of information is unclear, it should not be divulged.

- Someone who speaks on a controversial subject should not be quoted without permission.

- Electronic mail should be treated as privileged in the same manner as first class U. S. mail.

- Closed bulletin boards or conferences are private; however, participants in these activities should be aware that eavesdroppers who have not subscribed to any confidentiality agreement may feel free to use any information they can gain as they see fit.

- Open bulletin boards are public; they are analogous to physical bulletin boards, letters to newspaper editors, and postcards.

- Inadequate protection of confidentiality should be exposed if the personal rights of others are at stake and if the proper internal channels for protest and disclosure have been exhausted.

- Governments should go through proper channels to retrieve information from corporations, although clandestine acquisition of information about private organizations may be justified in some circumstances.

- In obtaining information, governments should not compromise the ethics of their citizens.

- In the absence of a legitimately documented need in the interests of national security, a company and its employees are within ethical bounds to place whatever safeguards they deem necessary on access to private company information.

Use of Information and Informed Consent

In today's information society, the issues surrounding the use and misuse of information have become increasingly important. Data bases are proliferating in both the public and private sector. Existing data bases are being merged, often with specious legal arguments as justification. The usefulness of having comprehensive information about a subject available in one computer file does not excuse unethical behavior, however. The release of personal and sensitive information on individuals available in government and other data bases must be limited. In addition:

- Use of personal information, voluntarily provided, for purposes other than agreed to, is unethical.

- By accepting and using various services, individuals are voluntarily releasing information about their affairs to others. They retain the right to inspect and correct information maintained in service records, however.

- Use of publicly available, legitimately acquired information is not unethical.

- Experimenting with human beings without their permission is unethical.

- Human subjects also have the right to withdraw from any experiment, for any reason, whether or not they initially gave informed consent.

Property or Innovation

In addition to physical property, individuals also own their names and the use of their phone numbers. Using people's names without their direct or implied permission is wrong, as is calling a phone number unless one is actually trying to contact that person. All property, whether physical or not, should be respected and used with care. Stealing, for whatever reason, is not ethical. Furthermore, the existence of temptation does not justify irresponsible action (e.g., the absence of security on a computer system does not justify illegal entry any more than the absence of locks justifies theft). More specific principles relating to property include the following:

- The knowing use of illegally acquired property is as unethical as the act of illegally obtaining it.

- Independent development of a software product performing the same functions as an existing product is a new product even when the function of the new product is similar to the original. (This assumes that no one else's copyrighted property, such as source code, was copied.) As a practical matter, current case law applies much more stringent requirements for products to qualify as new, not derivative, products. Thus, such software development may not constitute unethical behavior, but may prove to be illegal.

- Public formats, conventions, and standards are not copyrightable any more than the general shapes of other products, such as cars or coffeemakers.

- Because innovation and improvements to existing products should be encouraged for the good of society, individuals should be rewarded for being innovative.

- Private individuals and companies are not ethically bound to distribute their resources equitably. Public agencies are.

Harm

Any action, undertaken with malice, is unethical, as is harming others needlessly and damaging property by neglecting to exercise due diligence.

Use of Computers and Other Devices

Although a device is not unethical, certain applications of a device may be. Moreover, replacing a manual activity with an automated device does not change the ethicality of the activity. Using a computer as a marketing tool, for example, is not unethical. When the device being used is hidden, however, ethicality becomes more difficult to judge. A hidden device used in an accepted relationship is unethical, but enhancing one's intellect with a device in game playing, or elsewhere, as long as such use does not violate commonly understood rules, is ethical. Given our increasing technological capability to develop ever more powerful miniaturized devices, the controversy on this issue can be expected to increase.

PROFESSIONALS

Professionals, who include consultants, teachers, and managers, are expected to observe the rules of ethical conduct, as well as know and abide by relevant laws and contractual agreements. Although intent is a necessary component of unethical actions, professionals rarely can legitimately claim that they harmed someone unintentionally; they are expected to know what is ethical and what is not.

Responsibility

Professionals have a responsibility to do good work and to be accountable for that work. This obligation includes taking whatever steps are necessary to remove any impediments to properly performing the job, including bringing problems to the attention of a higher authority, protesting unreasonable constraints, making problems public (whistle blowing), refusing contracts, or even resigning from a position. The lengths to which professionals must go to maintain their ethical standards depend on their certainty of the problem's existence and the severity of the expected damage if the inadequacies are not corrected. Professionals must accept responsibility for prudently pursuing solutions for the higher good.

Students should observe professional ethics during their education, although most errors should be forgiven and sanctions limited. Students often overestimate the level of their knowledge and ability. Until their skills have been sharpened, they should not be judged with the same severity as practicing professionals.

Where significant technical expertise is needed, those with that expertise must be responsible for that work and be held accountable for the results. For example, although engineers may be responsible for designing the specifications for computer programs, analysts and programmers are responsible for the programs they develop according to those specifications. Authority and responsibility should be explicitly and carefully assigned among all participants in a task. Making such assignments is one of the most difficult problems in law or ethics. Disputes over the apportionment of responsibility are a sign of the maturing of the computer profession relative to other long-standing professions.

Protection of Rights and Interests

Scholars and other professionals have a right to determine when their work is ready for distribution. This right should not be unilaterally preempted, either by employers or by other professionals. Computer programs should not be distributed without the developer's consent, for example.

Professionals should not invite others to violate ethical principles. Encouraging users to attempt to compromise a computer service, as a way of trying to unearth system vulnerabilities, leads to situations inviting unethical behavior. Such questionable actions (except in a self-contained laboratory setting intended for that purpose) are dangerous.

Self-Professional ethics cannot and are not intended to protect individuals from the consequences of their own business decisions. Rather, they are intended as a standard of behavior against which professionals can measure their actions. Professionals thus have a responsibility to safeguard their own interests. In submitting a proposal, for example, they should indicate any proprietary interest and avoid including so much detail that someone else can do the work using their methodology. Contracts should ensure adequate current and future compensation. In the case of an expert system, this might mean receiving a percentage of the profits or a royalty for each package sold.

Public, Community, or Country-Professionals have a duty to avoid harming the public. They are responsible for the foreseeable consequences of the public's use of their products.

Client or Customer-Professionals are also obliged to offer services in the best interests of their customers. If they know an assignment will produce an inferior result or fail to accomplish the customer's purpose, it is unethical to solicit or accept that assignment.

Misrepresentation and Misleading Others

Professionals have public stature and thus a responsibility to avoid misleading the public, either by intentional misrepresentation of a product and its performance, or by attempts to mislead or deceive people. This responsibility also applies to more subtle situations, such as agreeing to perform research on a project they doubt will succeed. Their acceptance of the funding implies that they believe the research is worthwhile and can be successfully accomplished.

Professionals should also frame their public pronouncements carefully. Because they are regarded as experts, they should try to ensure that their meaning is not skewed if they are quoted out of context.

Professionals are responsible for ensuring that their products perform as promised. If a manufacturer consults with a machine designer on a machine to be made to certain specifications, the designer is responsible if the machine does not perform according to the claims made for it. Software tools, like hardware tools, should also perform as promised.

Credit

Professionals must always acknowledge their use of others' ideas in their written work for two reasons. First, misrepresenting derivative work as original is dishonest and unprofessional; second, authors of scholarly works have an obligation to their readers to help them find related studies. Therefore, quoted material should be so identified and fully referenced. Giving excess credit is always better than failing to give credit where it is due.

Conflict of Interest

Individuals or companies have the right to sell their expertise to more than one buyer. However, consultants' performing similar work for two competing clients is a conflict of interest, unless the work is done with both clients' knowledge and approval. Submitting a bid for work that creates a conflict of interest is unethical.

Contracts

Professional ethics also require that professionals deliver the product agreed to in the contract; therefore, professionals should not agree to perform work that they know cannot be successfully completed. Withholding a part of the expertise that they had contracted to provide would also be unethical.

When national security is at stake, governments should legally restrict the work citizens can do for foreign clients. Professionals are thus obliged to be informed of all relevant conditions and restrictions before making commitments to perform work for foreign clients. However, when no law prohibits the acceptance of a contract to perform such work, consultants are ethically bound to observe the contracts they have signed. This obligation includes protecting client proprietary information, even when sought by their government's representatives. Ethically, a government must use due process to obtain company information and should not expect professionals to breach their contractual responsibilities by providing data without their clients' knowledge and consent.

Use of Information and Informed Consent or Permission

Collection of customer information should be limited to the current information necessary for business purposes. Those whose records contain data on individuals' personal habits have a duty to protect that information. Collecting an excessive, and unnecessary, amount of personal information about individuals is unethical.

If professionals need to use personal information for extraordinary purposes other than those for which it was originally collected, they are ethically obligated to obtain the informed consent of the individuals involved. To act in the best interests of their clients and customers, professionals must insist that even police and other government authorities obtain the proper legal authorization to gain access to sensitive records. On the other hand, government authorities are not acting unethically in simply asking for access. Public servants should seek the least costly, legal way of obtaining essential and material information without violating ethical principles.

Programmers and systems analysts should always seek direct and positive authorization for the use of data files from whomever is identified through their best efforts as the custodian of those files.

Property

When computer programmers develop software tools that increase their productivity, they are entitled to continue to use them when they move to a new employer. Just as mechanics own their tools although they use their employers' machinery to sharpen and otherwise maintain them, so programmers who develop software tools using their employers' computer facilities have the right to continue using those tools.

Harm

Experiments involving human subjects should be well designed and stopped if deleterious effects are seen. To the extent possible, these negative effects should be reversed.

EMPLOYERS

Responsibility

Managers are expected to know what is happening in their organizations. Ignorance, or even avoiding knowledge, can mitigate culpability somewhat, but doing only what one is told is never an acceptable excuse. Furthermore, the size of a manager's organization is not a criterion for determining or limiting that manager's responsibility for subordinates' actions and performance.

Unethical acts by managers and owners in the name of their companies that become accepted industry practice are not thereby made ethical. Unethical acts are just as wrong whether done by 1 or 1,000.

Just as individuals are responsible for protecting their own interests, so too are employers responsible to exercise due diligence to protect their assets. As the agents of the employer, managers are required to act in the employer's best interests; that is, if they become aware of suspicious activities that may be unethical or otherwise detrimental to the employer, they must resolve those suspicions. To do otherwise would be unethical.

While employees may feel constrained to break an employer's rules on the basis of their ethical judgment, employers may be within their legal and ethical rights to impose sanctions on employees for breaking those rules.

Privacy

An employer has the right to determine how the organization's premises and equipment will be used, including restricting the employees' personal use of E-mail, other intra-office communications, and computer services. If such a policy exists, the employer is entitled to enforce that policy by monitoring the employees' use of these company resources. The policy must be consistently (not selectively) enforced. Otherwise, managers randomly searching employee files, whether electronic or paper, is unethical.

Similarly, when managers detect explicit wrongdoing, they have the right to monitor employees' activities specifically related to that suspicious activity. However, workplace surveillance in the absence of such a policy or cause for suspicion (e.g., for general purposes of detecting employee dissatisfaction) is an invasion of privacy and unethical. If a company wishes to routinely monitor employees, the employees must be so informed. They need not consent however. If the policy is unacceptable, they may choose to seek other employment or make it an issue in contract negotiations.

Other privacy issues that were raised during the workshop included:

- Personal habits of employees become a legitimate employer concern only when they interfere with job performance.

Otherwise, employers have no right to concern themselves with their employees' personal lives.

• Giving employees or prospective employees tests designed to reveal character without informing them of the purpose of the tests and how they are to be used is an invasion of privacy and therefore unethical.

• Personnel files are confidential except when a public entity properly authorizes their disclosure.

Property

Computer services managers have a professional obligation to discourage users from treating the computer irresponsibly, as well as to discourage casual and cynical attitudes toward computer usage. Allowing such attitudes to develop is dangerous.

Determining ownership of an asset such as a computer program developed using varying degrees of the employers' and employees' resources can be difficult. Development of solutions to this problem is increasingly critical.

Harm

Administering a test to employees that employers or their agents know may not produce accurate results is unethical, because of the possibility of harming those who are tested.

Policy

Management has the right to make policy for the organization and to expect staff to obey that policy unless doing so would be unethical or illegal. At the same time, when that policy directly affects their employees, employees must be made aware of the policy and other relevant terms of their employment. The policy should consist of general enforceable principles rather than describe narrow, specific acceptable or unacceptable actions. Managers have the right to determine the consequences of policy violations and to enforce those consequences within the limits of employment, as long as both the policy and the consequences are made clear to the employees.

Fairness and Equity in Evaluations

Just as they may monitor employees under limited circumstances, employers also have the right to monitor their employees to evaluate their job performance. However, using only one source of data would be unfair and unethical when other valid measures are available. To act ethically, employers must apply policies fairly to all employees. Moreover, employees have the right to know the criteria being used to evaluate their performance.

EMPLOYEES

Responsibility

In accepting employment, employees implicitly, if not explicitly, agree to observe company policy as a condition of that employment. Having agreed to abide by specified conditions of employment, employees are ethically obligated to abide by those regulations. If in doubt as to whether they apply in certain circumstances, employees should clarify the situation before acting. They should recognize and observe the intent as well as the letter of all legitimate company policies. Employees should not observe rules that are unethical or illegal.

As employees, if they believe that their management is treating them unfairly, they should not retaliate by shortchanging their employer's customers. Such actions on the part of professionals are unethical. The professional, as an employee, is obliged to seek a balance in serving the best interests of his employer on one side and customers on the other.

Employees are ethically obligated to protect their employer's interests and reputation to the extent that those interests are not unethical. For example, computer personnel have an ethical responsibility to inform the appropriate authorities when suspicious of computer misuse.

Privacy

Employees have a right to privacy. Employees browsing through information that is clearly personal or private, that does not belong to them, and that they have not been authorized to access is unethical. Employees should be aware when using their company's E-mail that it may be more accessible and more open to others' perusal than other intraoffice mail. Therefore, E-mail should be used with circumspection.

Property

Unauthorized use of any company resources, such as computer service or E-mail, may be illegal. Therefore, nonemployees, whether previously employed by a company or not, should obtain explicit authorization to enter company premises either physically or electronically. Under any other circumstances, the absence of authorization is implicit, and the entry is unethical and probably illegal.

Policy

If employees disagree with a company policy, they should try to have it changed. Until it is changed, however, they should continue to observe the policy.

Outside Business Projects

With the employer's full knowledge and agreement to the arrangement, and taking proper measures to protect against potential problems and conflicts of interest, it is not unethical for employees to pursue outside business projects.

Appendix

Position Papers

General Comments on Workshop

John McLeod, P.E.

The Workshop was an extremely interesting and — I believe — worthwhile exercise. It is not possible to legislate ethical behavior; we must, therefore, rely on education if we are to improve human behavior. And the Workshop was an educational experience, not only for participants, but also, perhaps, for a wider audience. There will be a report and probably, as in the case of the earlier SRI meeting, a book which we can hope will have wide distribution.

However, in considering the scenarios mailed to prospective participants prior to the Workshop it seemed to me that we would be "treating the symptoms rather than the disease"; the scenarios were concerned with the symptoms, the disease being a lack of uniform guidelines for ethical behavior applicable to ANY dilemma involving ethics.

So I made an attempt to draw up a list of "Generic Questions to Help Determine the Ethicacy of an Action," and sent it to Bruce Baker for comment. When I got to SRI I was pleased, and surprised, to find that Bruce had made copies of my effort for distribution to all Workshop participants. I didn't get as much feedback as I would have liked, but combined with other comments pertinent to the subject made at the Workshop, I have reworked my thoughts as follows.

* * *

IS IT ETHICAL?

TO BE ETHICAL AN ACTION SHOULD ELICIT A POSITIVE RESPONSE TO ALL APPLICABLE PRIMARY QUESTIONS BELOW, AND A NEGATIVE RESPONSE TO EACH CLARIFICATION THAT FOLLOWS THE QUESTION.

- IS IT HONORABLE?

  Is there anyone from whom you would like to hide the action?

- IS IT HONEST?

  Does it violate any agreement, actual or implied, or otherwise betray a trust?

- DOES IT AVOID THE POSSIBILITY OF A CONFLICT OF INTEREST?

  Are there other considerations that might bias your judgment?

- IS IT WITHIN YOUR AREA OF COMPETENCE?

  Is it possible that your best effort will not be adequate?

- IS IT FAIR?

  Is it detrimental to the legitimate interests of others?

- IS IT CONSIDERATE⸮

  Will it violate confidentiality or privacy, or otherwise harm anyone or anything⸮

- IS IT CONSERVATIVE⸮

  Does it unnecessarily squander time or other valuable resources⸮

CONSIDER THE FACT THAT ETHICAL SUPPORT OF UNETHICAL BEHAVIOR OF OTHERS IS SUSPECT.

Discussions during the Workshop underscored a fact that we all know; it isn't always practical to be ethical. There may well be overriding considerations. For instance, an employee's personal ethics might dictate that he "blow the whistle" concerning some unethical behavior of his employer. However, if that would cost him his job, and make chances of his getting another good one more difficult, his obligation to support his family might well override his ethics.

This led to a discussion of a hierarchy of ethics. Some actions may be more ethical than others! For instance, it is unethical for an employee to do anything contrary to the terms of his employment. It is also unethical to be dishonest. If after being given the facts, an employer directs an employee to make promises to a customer that the employee knows he cannot fulfill, what is the ethical (not necessarily the practical) thing for the employee to do⸮

In cases where a hierarchy of ethics is involved, the individual must rely on his own sense of values. Which is more valuable to him, his allegiance to his employer or his personal integrity⸮

I believe that ethical guidelines can be conducive to better behavior by individuals, and that in turn makes for a smoother working and pleasanter society in which to live. However, ethical guidelines must be complemented by a sense of values acceptable to that society.

* * *

SCENARIO II

SOFTWARE COMPANY: IGNORING VOTING MACHINE MALFUNCTIONS

Jim Moor

Opinion about this scenario may be vary in part by how closely the software company, XYZ, is understood to be allied with the overall project of supplying accurate voting machines. One could argue, as Smith's superior apparently does, that XYZ has done its job and it's not XYZ's problem that some of the machines are likely to malfunction. A tempting analogy is that a manufacturer of general software is not at fault if some of its software fails to run correctly because a random hardware malfunction. However, this does not seem to be the right analogy, for in this scenario the software is specially designed for the voting machines and the hardware breakdown is predictable. Whether XYZ likes it or not, its reputation may be damaged if these voting machines are discovered to be defective. Future sales of software for such voting machines is more likely if the voting machines perform well. In short, it is in XYZ's self-interest to promote the overall accuracy of the voting machines even though they don't manufacture the machines. Responsible behavior and good business practice are not inconsistent.

Of course, there is a larger social issue in the scenario which transcends the business prospects of XYZ. Fairness in voting is crucial in a democracy, and any dilution of such fairness undercuts the democratic basis for political expression and power. The scenario doesn't make it clear whether the malfunctions will be easily detected or how bad they will be. If major election results were invisibly altered through the internal operations of the voting machines, then the damage would be particularly insidious. Software developers are in a privileged position to detect and prevent certain kinds of computer malfunctions even when they do not cause them themselves. Software developers have a special responsibility to their clients and society to deliver quality products.

SCENARIO III.1

COMPUTER HACKER ("BREAKER"): ACCESSING COMMERCIAL COMPUTER SERVICES

Danielle Pouliot

A.1   Professional ethics is a term which must be seen exclusively from a context of professional behavior within the professions. In this sense, the student's behavior cannot be judged on the same scale as professional ethics. In this case rather, traditional concepts of right and wrong, moral and amoral, are more appropriate.

A.2   Considering our position in A.1, the behavior cited in A.2 and A.3 is not considered relevant to professional ethics.

SCENARIO III.1

COMPUTER HACKER ("BREAKER"): ACCESSING COMMERCIAL COMPUTER SERVICES

Danielle Pouliot

A.1   It is not only unethical but illegal as the interception of computer functions is illegal under the Criminal Code of Canada in 1985.

A.2   Even then, not only it is unethical but illegal. Obtaining computer services without requesting the right is an illegal act under the Criminal Code of Canada.

A.3   It is not a question of ethics. We believe the act is secondary to camouflage the illegal act cited in 1 and 2.

B.   Data processing systems are considered private property and people are responsible for respecting this property. Case history indicates that the young hacker did not act with malice, however, for me, the attempt to penetrate the company's computer system can only be done with malice. It is the same as trying to break into a home by using different keys, the act itself of trying to enter is a malicious act. The fact that the young man was invited, once in contact with the computer system, to use a service of the bank, does not legitimize what he did because he was able to access this invitation by an act which is malicious. It seems apparent that the bank did not have perhaps the necessary security measures to counter these types of activities, however, I believe that the failure in the bank's security does not excuse the behavior of the young man. It is like saying that because a house is not sufficiently secured, it is justifiable to break in. It is a question of responsibility, as I stated before, the responsibility to respect private property such as computer systems and I believe that the young man lacked integrity, I believe it is dishonest to try to penetrate a company's computer systems. There was, however, a lack of responsibility on the part of the company, to have sufficient security.

C.   Computer systems are private property like a car or a house. We have the same responsibility to respect this private property.

SCENARIO III.1

COMPUTER HACKER ("BREAKER"): ACCESSING COMMERCIAL COMPUTER SERVICES

John W. Snapper

Four acts are questionable: (1) the telephone scan, (2) the entry into the service, (3) the hacker's use of someone else's name, (4) the bank's claim against the hacker. Interestingly, (1) and (3) are paradigm examples of privacy violations according to Prosser's influential study (Prosser, The Law of Torts, #117).

(1)   A telephone scan is improper in so far as it disturbs the peace of telephone owners without justification in terms of the owners' interests. In the extreme, the hacker may be ringing phones in private homes. This is akin to nuisance calls by social pests, and worse than telephone advertising campaigns that are partially justified by the potential interest of the telephone owner in the advertised product.

(2)   Entry into an open service is permissible. Not only was there no attempt by the bank to prevent entry, the door was open with a public invitation to enter.

(3)   Assuming that the hacker did not have permission to use the name, the name is misappropriated. It is irrelevant that no negative reputation attaches to the owner of the name. I am harmed if my name is added to the faculty list at Harvard University or to the membership list of MENSA, although I would be flattered by the former and insulted by the latter affiliation. In either case, I am harmed by loss of control over my name. In addition the backer has apparently lied. If he had used a fictional name, he would still be intentionally misleading the bank. This is an unthankful response to those nice bankers who have offered him free service.

(4)   The only complaint that the bank may have is over the false use of a name, not over the use of the wide-open service. Under these conditions, the false name is an inconvenience and not a serious harm to the bank. Since banks carry great weight in legal and social confrontations, it seems out of proportion for the bank to either threaten legal action or to make public accusations.

SCENARIO III.2

CONSULTANT AND EX-MANAGER, COMPUTER FACILITY: DEMONSTRATING INSECURE COMPUTER OPERATION

Danielle Pouliot

A.1   Many scenarios deal with the case of a person who breaks into a system to prove that it is not foolproof (for example, case 3.3).

I noticed that, in general, the experts who attended the workshop did not consider this conduct to be unethical. In this particular case (2.1), firstly, because of the unauthorized use of data concerning that person and since it was, as far as they knew, the only means to become aware of the system's faults.

I believe that the judgments on the cases presented during the workshop should be based on the moral value of the behavior observed and not on the principle that "the end justifies the means".

I consider that the consultant's behavior is completely unethical for the following reasons:

- the consultant did not have his password invalidated as he knew he should have;

- the consultant used an unauthorized password;

- the consultant used an information processing system that was not authorized for his use;

- the consultant illegally obtained information which belonged to a third party;

- the consultant used unfair tactics in order to secure a contract in computer security.

First, the fact that the consultant accessed unauthorized data only to prove to the company that the security system was faulty does not justify his behavior.

At no time does the end justifies the means: the consultant's behavior is unethical and illegal.

Secondly, when a person is concerned with the integrity and integrality of systems and equipment, this person does not take measures to abuse these resources or to put them in danger; the contrary is entirely paradoxical.

Thirdly, by breaking systems' security the consultant may be seen as a model, and therefore perpetuate this type of behavior, instead of finding an alternate means of solving this problem.

The consultant's behavior was not only unethical, it demonstrated a lack of judgment, because he considered the problem as technical instead of managerial. The real problem is not the security of the system but rather the manager who is deliberately ignorant of the system's weaknesses. Because of the circumstances, the manager will make the necessary corrections to the system but will not be "sold" on system security, on the contrary. This can only be seen as a "band-aid" solution and does not solve the problem in the long term.

SCENARIO III.5

PROGRAMMER: PRODUCING NEW SOFTWARE BUILT ON AN EXISTING PROGRAM

John W. Snapper

The case contrasts the legal definition of software property with the extra-legal expectations of some software engineers.

There is the legal (rather than ethical) question whether the programmer infringed the copyright on the original product. It seems that he did not. The central tradition in property law is that copyrights protect the way a program is worded, not its function. There is no infringement when the new program is "different and independently produced," even if it performs similarly to the original. This conclusion is supported by the fact that no legal action was taken.

Against this legal background there are a number of ethical questions about (1) whether the limitations on copyrights that permit redevelopment are in keeping with our ordinary sense of private property, (2) whether copyrights are appropriate for computer software, and (3) whether the software industry accepts an extra-legal ethic that precludes redevelopment.

Much has been written on the first two questions. We cannot give them a fair treatment in this short discussion. (I personally do believe that the traditional limit on copyright protection is fully justified and must apply to software copyrights.) We can however draw some quick conclusions about the third question. There are many examples of plagiarism that are universally condemned, even though they involve no technical copyright infringement. But there seems to be no extra-legal notion of plagiarism in the software industry that is relevant to this case. Indeed, some software engineers at the SRI workshop expressed indignation at the idea that anyone would object to independent redevelopment of this sort.

The one unethical act here may be the threat of legal action by the original producer. When that producer copyrighted the program, he accepted the limits of copyright protection. There are then multiple reasons to condemn a threat of legal action in blatant disregard for those limits. Complaints about the means of protection should not take the form of accusations. In general threats need justification of a sort that is lacking here. Moreover, since these limitations on copyright protection are typically seen as preserving rights to independent intellectual research, the threat can be seen as an attempt to scare the independent programmer into giving up an important right.

SCENARIO IV.5

PROJECT MANAGER: DEVELOPING SMART CARD THAT STORES CREDITWORTHINESS DATA UNKNOWN TO CARDHOLDERS

Edwin B. Heinlein

On the assumption that the situation describes the granting of credit to a card holder based an information contained in the card and the location of which information the cardholder has no initial knowledge, I see no ethics issue.

The scenario details a smart credit card which obviates a merchant telephone call to a data base where cardholder creditworthiness information currently resides. It is the same information on which credit decisions are now made. In addition:

- Credit history data is no more an individual's property than his/her medical records. In the credit history arena, we have certain legally obtained rights to inspect and correct them.

- A credit/debit card is the property of the institution which issues it and the contract between the parties so states.

- Current credit and debit cards do not detail the content of the mag stripe nor is it necessary to do so. In fact, it seems to be a positive security matter not to publicize the stripe content or layout.

- The issuance and use of a credit or debit card is a commercial transaction between an individual and a financial institution. There is no requirement to enter any such arrangement if one does not wish to do so nor does one lose any rights or privileges if one does not do so.

The scenario states only that negative credit information was conveyed. Clearly, positive credit information was conveyed, as well, and probably more often since people with poor credit history are not the majority of cardholders.

SCENARIO IV.5

PROJECT MANAGER: DEVELOPING SMART CARD THAT STORES CREDITWORTHINESS DATA UNKNOWN TO CARDHOLDERS

Jim Moor

This scenario does not discuss the details of the information on the smart card. The scenario could be elaborated in such a way that the activity is unethical. Suppose the card stored detailed transactional information about purchases and other economic properties of the buyer, and this information was secretly passed to any store at which the card was presented. Such a situation would be an unjustified and needless invasion of the privacy of the individual cardholder. On the other hand, the scenario could be elaborated in less sinister ways. Suppose the card kept a running total of the card holder's balance on the card. At a point of sale the card would be inserted in a machine which would give a response of "approved" or "not approved" for an amount of a potential purchase. In effect, the card and machine would do what is often done today by calling a credit card company on the phone. In such a situation the card-holder would readily become aware of what the card did and no invasion of privacy would occur.

Like most ethical decisions the correctness of this decision depends a lot on the facts of the case. Different facts can justify different ethical conclusions. Several general ethical considerations are worth noting. First, telling cardholders what kind of information is contained on such a card would seem to be a minimum for truth in advertising the card. This does not mean that the consumer must be told every detail about how the card works any more then a consumer must be told the details of how a certain kind of medicine works. But, the consumer should be informed about what kind of results can be expected including what possible harmful side effects may occur. Second, in such a smart card system there should be a way for a cardholder easily to read the information on his or her own card. This is the best guarantee against errors and misunderstandings.

SCENARIO IV.8

UNIVERSITY STUDENT: OFFERING LIMITED ACCESS TO A PORNOGRAPHIC QUESTIONNAIRE

Jim Moor

The student who kept the pornographic questionnaire on the computer system was not causing harm. Nobody was forced to see the questionnaire, and since a warning was given, nobody was deceived about the contents of the file. The scenario does not suggest that any law was broken or that anybody's privacy was invaded or even that a university policy was abused. In short, the student did nothing wrong.

If the student did nothing wrong, then the Dean of Students' office was overreacting. Perhaps, the University officials believed they were protecting the morals of the students. Such efforts today are as misplaced as they were in the days of Socrates. The university officials may have felt some embarrassment because a university computer was being used to provide pornographic information or may have feared a decrease in alumni contributions if the alumni found out about it. Such embarrassment and fears are considerations but are not decisive. Universities, above all institutions, must encourage and protect free speech. University officials may have believed that because the university owned the machine, the university had the right to formulate policy. But, even if true, university policies about computers should be clearly announced in advance. In this kind of case, instituting a policy of censorship would be shortsighted. Whatever the short run gain, in the long term such censorship would weaken, not strengthen, values fundamental to universities and society as a whole.

SCENARIO IV.9

ATTORNEY: REQUESTING ACCESS TO COMPUTER SERVICE COMPANY RECORDS

John W. Snapper

A common reaction to this case is that the prosecutor should make no use of the records since there is no evidence that pornography has any connection to rape. But there are many ways in which the service company's records could be relevant to the prosecution, regardless of any supposed relation between pornography and rape. Transactional information on periods of use could establish that the defendant was using the service at the time of the rape and is innocent. If the rape was modeled on events described in a pornographic work (including how the rapist "named" the victim), it would be relevant to show that the defendant had been reading that particular work. Whether the prosecutor should have the records is largely a question of whether they are likely to contain information relevant to the prosecution.

Since the records may contain relevant information, it seems reasonable for the prosecutor to request them. This is an ordinary request, without ethical significance.

The records are the property of the computer service company. That company may agree to or refuse the prosecutor's request. In this situation, however, the service company has an ethical duty to refuse. Reading habits, particularly when they include sexual topics, are viewed in our society as a deeply personal matter. The service company therefore is holding information that relates to the private lives of its clients and should treat that information with care. It should certainly not open its records merely on request.

If the prosecutor still wants the information, he may seek a warrant or court order. In theory, an order would only be granted upon demonstration that the records are indeed likely to contain relevant information and would block the prosecutor from a mere "fishing operation." Alternatively, the defendant could authorize release of the information. The service company should then comply with the order.

In summary, the following series of events seems reasonable: a request by the prosecutor for the records, refusal by the service company, receipt of a court order, and compliance by the service company. The only act in this series that apparently involves ethical issues is the refusal by the service company, on the basis of concern for privacy, to comply with the original request.

There may be some ethical and legal debate on the grounds for the court order. Since the records contain sensitive personal information, perhaps the grounds should satisfy particularly high criteria. Perhaps the defendant should have a say on the order, even though the information is apparently the property of the service company. A full analysis of these issues is beyond the scope of this discussion.

SCENARIO V.1

CONSULTANT: PROPOSING AN INFERIOR COMPUTER PROGRAM

Edwin B. Heinlein

We found the situation hard to believe since companies which use consultants will have them execute confidentiality agreements to cover such situations. This led us to conclude that the consultant would have been more aware of this practice even if the companies were not. In the very least, we felt he should have informed one or both companies of his situation.

SCENARIO V.2

COMPUTER SCIENTIST: DIVERTING RESEARCH FUNDS TO ANOTHER PROJECT

John M. Carroll

I found the discussion of this case disturbing. I was appalled to hear scientists and philosophers invoking exculpatory nostrums such as:

"Everybody is doing it."
"Science is a more important calling than bookkeeping."
"The end justifies the means."
"Nothing wrong was done because nobody got hurt."
"Sloppy management is not an ethical issue."

When we resolve questions of right or wrong using contemporary behavior as our reference we have abandoned ethics in favor of ethos. It seems that our high priests of secular rectitude are measuring moral values with a rubber ruler.

If the salt of the earth has lost its savor, wherewith then shall the earth be salted?

The arguments in favor of improperly diverting funds in this instance admit of extension to legitimate all manner of disreputable conduct not the least of which are lying and stealing. Even a minor diversion is but one step down a slippery slope that leads only to perdition.

In the first instance, the scientist should have sought a reallocation of funds by higher authority. If such a reallocation was not forthcoming, the scientist should have done the best he or she could with the funds available.

Surely a temporary failure in one project would be less costly in the long run than relinquishing the moral high ground to embrace the ethics of dialectic materialism. On putting it another way: How much does it profit a person to gain the whole world at the cost of one's spirituality?

SCENARIO V.3

INFORMATION SECURITY MANAGER: PROTECTING MULTINATIONAL COMPANY DATA FROM U.S. GOVERNMENT ACCESS

David Burnham

"National security" claims, if valid, would justify the attempt of the US government to obtain information about the activities of the South Africa division of a company. The scenario gives us no indication whether there is a valid national security requirement. We have many recent examples, of course, when such claims appear to be fraudulent. I think the security manager can properly protect the privacy of his company against snooping.

SCENARIO V.4

SECURITY CONSULTANT: NOT DISCLOSING FOREIGN COMPANY DATA TO U.S. GOVERNMENT

David Burnham

In this case the government openly asks for information. In making the request, of course, the government would have to state its reasons. This does not seem improper. The decision of the security consultant could be influenced by whether he felt the government's claim was both legitimate and important. He still has a legal and ethical obligation, however, to his employer which in most situations would make it appropriate for him to protect the data.

SCENARIO V.4

SECURITY CONSULTANT: NOT DISCLOSING FOREIGN COMPANY DATA TO U.S. GOVERNMENT

Danielle Pouliot

A.1 The American Government goes against principles of professional ethics by requiring the consultant not to respect the confidentiality and nondisclosure clause for the contract with his Saudi client.

A.2 It is a demagogical argument which supports an act contrary to professional ethics.

A.3 The fact that the American Government is delayed in its efforts to obtain information on the company is a consequence of the consultant's activities and not an intention of the consultant to prejudice the American Government.

B. Apparently, there is no law in the United States which prohibits American consultants to carry out contracts in Saudi Arabia. In this sense, the consultant seems not to have lacked professional ethics by doing business with a Saudi company. Furthermore, it seems normal to us that the consultant includes a confidentiality and nondisclosure clause in his contract stating that he would handle confidential information belonging to his client; this is a normal procedure.

Furthermore, it is difficult to understand why, if the American Government needs to obtain information on the compa-

company, the mandate carried out by the consultant hinders it to do so. This gives us to understand that the American Government could access information to which it is not entitled and in a highly irregular way. In our opinion, the fact that the Saudi firm wants to protect its information should not prejudice the fact that the American Government could access the information on the company since the information, which it could access, should be publicly provided to it. If such is not the case, the American Government is trying to make the consultant an accomplice to a procedure highly disregarding professional ethics.

C.   Only the contract, which binds the consultant to his client, prevails. The consultant must respect the confidentiality and nondisclosure clause included in this contract.

D.   If the consultant had signed such a contract, knowing it would go against American law or policies, or knowing such a contract could endanger the security of the United States, he would have highly disregarded principles of professional ethics. In this case, the American Government would have been justified to ask him to derogate the confidentiality clause of his contract with the Saudi company.

SCENARIO V.7

PRESIDENT SOFTWARE DEVELOPMENT COMPANY: MARKETING A SOFTWARE PRODUCT KNOWN TO HAVE BUGS

Richard T. De George

The president of the company tries to justify his action by three arguments. One is a business argument that claims that speed is necessary to gain market share. The other two hinge on facts about computer software. One is that bugs are typical in software and no one can guarantee absolute freedom from bugs in any complex program. The second is that commercial software is constantly being improved and later versions issued. This is taken to show that all software developers sell less than perfect software, which they continually improve.

However, these justifications fail to show that the president's actions are ethically justifiable. The scenario states that the president knows the program has bugs and it implies that he assumes this will lead to errors in income tax filing for some users. The scenario does not state whether the program has been field tested, as is the industry custom and norm. If it has not, the president acts irresponsibly in marketing it before it is so tested. Sale of the product is fair only if the normal expectations of the buyer are met; otherwise there is deception. Software purchasers rightly expect known defects to have been removed and reasonable accuracy of the product for its stated purposes. If the president violates these expectations, the transaction is unfair to the buyer and user. If he has field tested the program, knows it is defective, fails to correct it, and markets it anyway, he similarly violates the fairness conditions for sale of the program.

The president claims no responsibility for harm done a user by use of the product for its intended purpose. Attempting to evade responsibility when he knows the product is defective is unethical unless he specifies the defects he knows the product has and warns that it may have others (if he has reason to believe this). Otherwise he sells the product under false pretenses. If, when the program leads to filing mistaken tax reports, he is held liable for the mistakes by the IRS (as the tax preparer) or if he otherwise pays the users for their trouble — and financial loss; he may consider such payments will be more than offset by the volume of sales he achieves by his timing. This good to him has to be weighed against the harm done his users. Although the harm done those who misfile consists primarily of paying the additional amount due, plus interest and perhaps a penalty, they also suffer some trauma and anxiety from their encounter with IRS.

In general one is ethically responsible for the foreseeable consequences of one's actions and simply disclaiming such responsibility does not relieve one of it. It is unethical to violate generally accepted industry standards without so notifying the potential buyer, since this violates the fairness conditions for commercial transactions. It is unethical to falsely advertise that a product can do more than it is known to be able to do. Whether this applies here is not clear.

In attempting to justify his action the president misrepresents or misinterprets the situation of the software industry. Although any complex program may (and probably does) have bugs, programs are not (and should not) generally be sold with known bugs that can be fixed. Nor are (or should) programs for general users be sold before they have been adequately tested and found to perform the task for which they are designed and sold. Similarly, the fact that later versions of programs constantly appear does not show that earlier ones are defective or do not do what they are sold to do. Later versions are usually improved or enhanced — i.e., they work more quickly, do something earlier versions did not do — as well as being fixed to remedy any (usually) arcane defects discovered (and discoverable only) through wide use. Hence the facts and standards of the industry are misrepresented by the president of the company and they do not justify his sale of the program — which is premature, violates fairness, and causes foreseeable (on his part) harm to users.

SCENARIO V.7

PRESIDENT SOFTWARE DEVELOPMENT COMPANY: MARKETING A SOFTWARE PRODUCT KNOWN TO HAVE BUGS

Edwin B. Heinlein

In order to make my position as clear as possible, I have assumed that the software underwent at least the minimum acceptable testing. For me, this includes beta testing of a reasonable degree which is controlled user environment testing.

Since this is standard industry practice without which a package would be too vulnerable to an early demise from quickly spread word of unusability, I feel it is a reasonable assumption.

There are three issues; the decision to market a product which probably has bugs, the disclaimer and the reason for the disclaimer.

My position is that no software system of any size is ever bug free. What reasonable initial testing comprises is stated above. The depth and extent of the testing is a balance between making a market window opportunity and having a product which is usable. My experience is that bugs will show up in software even after many years of use due to the building and combining of conditions previously not encountered.

Disclaimers are used an many, if not all, products in the market place. They are necessary to avoid the current legal interpretation of product liability. For instance, the large interior windshield sun screens have a notice printed on them warning against driving with the shield in place. Also, a manufacturer of small portable radios which play through a pair of earpieces is being sued for not having a warning in the package about using the product while crossing the street. Someone was injured while doing just that.

The president of the company has a duty to protect all interests of the company from liability issues. Those protected include the investors, management and the employees. The disclaimer is the legally accepted method to do this. That it happens to be industry practice is stating a fact and not a reason.

I see no ethics issues with any of the actions and, in fact, would consider the company executive irresponsible if the product were marketed without the disclaimer.

SCENARIO V.8

MARKETING MANAGER: DEVELOPING QUESTIONABLE TV COMMERCIALS

John M. Carroll

The essence of this case is the use by a manufacturer of personal computers of advertising that might mislead some potential buyers.

There is no suggestion that the advertising was false in any legal sense, neither does the advertising make any promise that might create an implied warranty.

Why then should promoters of computer sales be held to higher standards of conduct than normal business ethics, something cynics regard as an oxymoron anyway?

One cannot argue the magnitude of the selling price demands higher standards because today that price can be much less than that of a used car.

Nor can one argue public safety. Nobody has yet died because of an unsafe computer. (We exclude terminal boredom.)

The most persuasive ethical argument against this kind of advertising derives from the concept of fairness: A business transaction should be fair to both parties. A transaction cannot be said to be fair if a great disparity exists between the information resources available to each of the parties.

The possessor of the greater store of specialized knowledge enjoys a privilege and the principle of noblesse oblige dictates that privilege incurs obligation.

In this case, the obligation is to be candid with the other party. Here candor should go beyond mere truth-telling and approach complete disclosure.

As a sidelight, we posit the proposition that it is not only the naive consumer, who is hurt by mindless and misleading advertising. Other manufacturers who promote their wares with candor lose both sales and credibility because of it. They lose sales because customers are attracted to the vendor who offers the most. Then, when things go sour all vendors tend to be stigmatized even the honest ones. Thus does unfair dealing meld into unfair competition.

Metaphoric television advertising that implicitly overstates the capability of a complex product inadequately understood by the

targeted consumers may be acceptable as an atte
should, however, be followed at the time of sale by
sion of the products capabilities and limitations. T
to ensure when sales are made by dealers who have an arm's length
relationship with the manufacturer.

SCENARIO V.9

CONSULTANT: PERFORMING PRO BONO PROFESSIONAL SERVICES

Danielle Pouliot

A.1   ... insofar as his customers agree.

A.2   It goes against professional ethics to publicly discredit the methods or decisions of a colleague all the more so since they did not prejudice this profession.

B.   Insofar as the consultant does voluntary work and does not try to profit from this situation, he does not go against any principle of professional ethics, same as would a dentist offering free services to underprivileged families.

SCENARIO VI.5

RESEARCH CHEMIST: USING BULLETIN BOARD SYSTEM (BBS) FOR RECRUITING

Edwin B. Heinlein

Other than me, none of the group had encountered this issue previously and I added what I knew; that such policies do not restrict the individual from making an advance nor the company from replying. These policies are a form of truce when otherwise competing companies get together for a mutually beneficial project and have always been temporary, i.e., for the life of a project. Our discussion hinged an the matter of the policy being poorly stated, that the U.S. Mail was specifically called out and that this invited hair splitting in the interpretation. Also, we concluded that it would not be possible for the chemist and the recruiter to be unaware of the intent of the policy.

SCENARIO VI.5

RESEARCH CHEMIST: USING BULLETIN BOARD SYSTEM (BBS) FOR RECRUITING

Danielle Pouliot

A.1 The person in charge of hiring should have been aware this practice went against the company's policies.

A.2 It is not a question of ethics. The company has the right to issue policies it considers appropriate.

B. Even if the use of electronic bulletins was not directly company policy, the chemists should have known that it was forbidden to use the regular mail for hiring directly other chemists.

The fact that electronic bulletins not being company policy does not justify the chemists to benefit from the flaw, since the principle was already clearly established by forbidding the use of the regular mail.

SCENARIO VI.6

INFORMATION SECURITY MANGER: MONITORING ELECTRONIC MAIL

David Burnham

A company has the right to establish standards concerning the use of electronic mail system. The scenario does not tell us whether there were such standards and whether the employees had been told about them. The scenario also does not make clear whether the love letters, homosexual liaisons, etc. had any impact on job performance. If they did not, the company is on ethically weak grounds to use information to discipline employees.

The New York Times, of course, has a very elaborate word processing system. Several years ago, someone in the Times used the system to intercept an angry exchange of messages between the executive editor, Abe Rosenthal, and one of the paper's best foreign correspondents. This exchange subsequently was printed in the Village Voice. Rosenthal was furious. Three weeks later, a memo was distributed to the staff which set down rules for using the system; book writing was forbidden, game playing was forbidden, electronic eavesdropping was forbidden. The memo began with a grand phrase that professionals do not intercept the messages of other professionals. It ended with the announcement that the NYT would now begin inspecting the "directories" of individual reporters and editors to make sure the rules were followed. It was okay for the NYT to secretly inspect a reporter's electronic file, but not for a reporter to look at the files of the NYT.

I was a NYT reporter when this memo was circulated. It caused almost no discussion. I thought at the time that if the NYT had announced it was going to begin regularly inspecting the desk drawers of individual reporters to make sure there were no personal materials stored in them, that there would have been a good deal of resistance. The word processing system, however, was sufficiently abstract so that it didn't hit home. Is there a generality here that the ethical issues raised by the computer are inherently more difficult to deal with than the issues raised by a physical search of a desk drawer? I think so.

SCENARIO VI.6

INFORMATION SECURITY MANAGER: MONITORING ELECTRONIC MAIL

Richard T. De George

This scenario raises a number of issues. One is the status of a company's electronic mail system. In this scenario we assume the company has not issued any rules clarifying and governing appropriate use, privacy, and similar issues. The information security manager routinely monitors the contents of the system. The scenario does not indicate why the security manager monitors the system, but the implication is that security is an issue and monitoring helps preserve it. Since the system belongs to the company and is primarily, if not exclusively, for company business, the company has the right to see to its efficient and proper (as it defines this) use. This right comes from its ownership of the system.

The security manager found that a number of employees were using the system for personal purposes. He routinely informed the human resources department and corporate security officer about these communications. The fact that he informed these persons about personal use of the system by some personnel is ethically permissible. It is here, however, that an ethics issue arises. For he not only reports that the system is being used for personal purposes but he supplies printed listings of the personal messages. The personnel who used the system for personal messages may have acted unwisely in so doing, since they know the system can be (and if security is an issue, probably is) monitored. In the absence of any rule prohibiting personal use, they do not act unethically, since the company's telephone system and interoffice mail system have (the scenario implies) been similarly used in the past and such use has been allowed. Two differences are that these, we assume, were not monitored, and that no permanent record was available or kept.

Despite the fact that the security manager did not act unethically in monitoring (since that is part of his job) or in reporting personal use in general, the manager did act unethically in providing others with detailed contents of clearly personal messages. No security purpose is served by providing the contents of these messages, nor is any other clear company purpose served. Hence the right to privacy of those sending the messages is the central ethical principle operative here. If the company managers, on learning of personal use (which they could easily have anticipated) wish for some reason to preclude it, they should so notify all employees. If they do not wish to preclude it, they should notify employees that the system is being monitored. It is only fair that people know the rules of the game in which they are involved. But if the managers either tacitly or explicitly allow personal use, ethically they must respect the privacy of their employees in their use of the system. The security officer may monitor the messages, otherwise the officer would not know if they violate security. But the officer has no right to copy, report the contents of, or in any other way act on the contents of personal messages that do not affect security. Although security considerations would override privacy considerations, in the absence of security threats, the employees' right to privacy should be respected. If the company prohibits personal use of the system, it should do so clearly.

Finally, the managers who punished employees on the basis of the contents of their personal messages acted unethically, if they punished them for non-job-related aspects of their personal lives (as the scenario implies). To do so exceeds the legitimate authority of the manager and violates the employees' right to privacy and to the freedom of action allowed all persons in their personal lives.

This case demonstrates the need for clear company policies regarding a company's electronic mail, since it is a new medium and since there are no clear shared assumptions about its proper use.

SCENARIO VI.6

INFORMATION SECURITY MANAGER: MONITORING ELECTRONIC MAIL

John W. Snapper

The Information Security Manager (ISM) is under some circumstances expected to read the EMAIL. If research is protected as a trade secret, for instance, the ISM may be required to review messages that pass through research areas, whether they be electronic or hard copy messages. But in this case, the ISM search went beyond permissible limits.

The ISM reported findings that did not bear on security. Assuming a trade secret environment, he should of course report messages that contain research results. But love letters do not bear on research and need not be reported. To take a mundane example, I expect librarians to search my papers as I leave a library and report any library materials, but would complain of a privacy invasion if it were reported that I carry a personal copy of Superman Comics for reading pleasure.

It may be relevant for the ISM to take note of fraternal relations between employees that pose a threat to security. Security leaks are often traced in this way. But it does not follow that love letters should be reported verbatim to management. The problem in this case may be that the ISM is also the access control administrator and thus may have additional responsibility to guard against non-work-related use of the EMAIL system. Although common, this combination of the two positions creates a potential for abuse of security checks.

Several aspects of this case suggest special privacy concerns: the dissemination of personal information, the violation of an expectation privacy, and the possibility that employment decisions are based on private matters.

There is no apparent justification for the ISM to disseminate listings such as love letters that include personal information. At most it is sufficient to inform the resource manager of the personal use of the EMAIL. Since betting information is less personal, we need not be quite so cautious with those listings.

If an expectation of privacy on the EMAIL has been created, it should be respected. Since the employees protested, that expectation was apparently created.

The possession of listings by the resource manager gives the impression that employment decisions are based on the specific content of those listings, rather than that the EMAIL was misused. That would be improper.

SCENARIO VI.6

INFORMATION SECURITY MANAGER: MONITORING ELECTRONIC MAIL

Danielle Pouliot

A.1   It is an unethical act, not really the fact of using electronic mail for personal use which is unethical, it is the manner in which the agent responsible for the computer safety did so. That is to say that he should have respected the confidentiality of the persons concerned in the use of electronic mail for personal use.

A.2   The employees did not respect professional ethics when they used a corporate resource for personal purposes.

B.   The information security manager went against the basic principle of confidentiality by disclosing the names of those using the corporate electronic mail for personal purposes. The person in charge of security did not have to provide the names of the individuals involved to prove that there had been personal use of the electronic mail. In fact, he simply had to furnish a copy of the communications. Moreover, the employees did not respect the basic principles of confidentiality by making personal use of a communications equipment which may be considered as public. Furthermore, the employees made personal use of a corporate resource which shows, in our opinion, a great irresponsibility on their part.

Finally, what we consider as being the most important is that the employees should have been notified that the information security manager made regular routine checks of the contents of communications on the electronic mail. These people should have been notified, in our opinion, that the electronic mail could be regularly verified.

C.   Professional ethics call for a rational use of each and every corporate resource.

SCENARIO VI.6

INFORMATION SECURITY MANAGER: MONITORING ELECTRONIC MAIL

Danielle Pouliot

A.1   The workshop clearly demonstrated the importance of businesses to clearly outline their expectations concerning employees. Often in order to judge the ethical nature of a behavior, we must know if the person concerned was aware that the rule existed.

The information security manager acted unethically because he intercepted employee electronic mail without warning them of these routine checks. What is more, he never clearly stated that it was prohibited to use electronic mail in this fashion.

SCENARIO VI.7

EMPLOYER: MONITORING AN INFORMATION WORKER'S COMPUTER USAGE

Deborah G. Johnson

Management has behaved unethically in this matter. in asking the security department to monitor the computer services activities of the worker, management has unjustifiably violated the privacy of the worker.

The issue raised by this scenario centers on the relationship between an employer and an employee, and the rights of both parties. The employer-employee relationship is an interesting area for ethical analysis because this area of our lives tends not to be well defined. That is, we tend to rely more on custom than on law when it comes to thinking about employer and employee rights and responsibilities. We all generally agree that it is wrong for someone to sneak into another's home, hide in a closet, and listen to everything that goes on. We also think it wrong for a person without a warrant to eavesdrop on another from outside their home, e.g., by wiretapping their phone. We even seem generally to agree that it is wrong for an unauthorized person to intentionally gain access to, and examine the contents of, another person's computer file. However, these activities are somehow transformed when they take place in workplace.

Only in special cases do employers specify the privacy status of actions, conversations, mail, writing, etc. done at work. Many workers go through the day thinking of their office at work as theirs; of the phone calls they make at work as private; of mail sent through the internal mail service as private; etc. In fact, this may not be legally so. Employers can make a strong case for a right to control everything in the workplace. After all, the employer owns the premises and all the offices, and the equipment in them. The employer pays the phone bills. The internal mail service is private and, therefore, not covered by federal laws. Indeed, all employees are, in some sense, agents of the employer.

This might seem to support the claim that an employer (or management in the scenario under consideration) has the right to monitor the activities of his/her employees. After all, the employer pays for this activity. The problem is that this is rarely ever explained to employees. Though not specified, it appears that the worker in the scenario was wholly unaware that her activities might be monitored.

Does a company have an obligation to inform its employees that it is going to monitor their personal (or professional) use of computer services? It is true that the company in this scenario may not have thought about monitoring its employees until the recent trouble developed. And it is true that if management were to announce that it was about to monitor use, much of the information which management wanted to see might be destroyed.

Nevertheless, a principle of respect seems at issue here. The employer-employee relationship is essentially a contractual relationship. The employer agrees to provide a salary, benefits, certain work conditions, etc. in exchange for work. An employee agrees to do the work in exchange for salary, benefits, etc. Such arrangements are morally acceptable when, and only when, both parties to the agreement are accurately informed (about relevant factors) and freely choose to make the agreement. Some might argue that it is up to the employee to inquire in advance as to the privacy conditions on the job, but this seems unfair given that the employee is less likely to know what to ask about. The employer is in the best position to know what conditions in the workplace will be of significance to the employee.

The antithesis of respect for a person is to treat people merely as means to your own ends, denying people the opportunity to make a decision about their behavior which will affect their life. When employers monitor the activities of employees without informing them, they are taking advantage of the ambiguity in the situation. They are keeping information from the employee which they know to be highly relevant to the employee's choices (about whether and how to use the computer services for personal activities).

Management should have told all workers that while they were allowed to use the computer services for personal activities, these uses of the computer might be monitored at some time. Management violated the privacy of the information worker when it ordered the security department to monitor her use.

SCENARIO VI.8

COMPANY MANAGER: FORBIDDING COMPUTER GAMES

Calvin C. Gotlieb

The company or manager cannot be faulted for making their position clear on how resources put at the disposal of employees are to be used, and in particular, for stating that company computers should not be used for game playing.

More controversial are:

(1) whether random inspection of employees' computer files is acceptable;

(2) whether the presence of a game in a file is sufficient evidence of non compliance with company policy to justify punitive action.

(1)   It can be argued that electronic files should be treated like conventional files. These are ordinarily not subject to random inspection, and there may be a personal component which is regarded as belonging to the employee. But there is as yet no norm for electronic files. Computer programmers will know, and users should know, that electronic files by design are usually more open, and that it is rare for them to have the same degree of security that can be achieved with conventional files. Nevertheless, if electronic files are subject to random surveillance, clear notice of this possibility should be given in advance to their owners. Without such notice, random surveillance should be considered improper and unethical; with notice random inspection must be accepted as a working condition.

(2)   There are situations, like the possession of burglar's tools, that carry with them a presumption of guilt. Here, even use of the file space is a prohibited resource consumption. In the presence of an explicit ban, and notice of random inspection, the manager has a right to the action; but it should be recognized that strict enforcement will be part of the working environment, and that it can be detrimental to recruiting.

SCENARIO VI.9

AIRLINE EXECUTIVE: AUTOMATICALLY MONITORING THE PERFORMANCE OF AIRLINE TICKET AGENTS

David Burnham

The executive's action is not ethical, if, as it appears, he does not inform his employees of his monitoring project. People have a right to know about such systems, I think the German Supreme Court called it the "right to informational self determination." The use of the program to make personnel decisions to the exclusion of all other factors is unethical because it is impossible to measure many important qualities with statistical measures. It also may be unethical because the executive is damaging the stockholders by his short sighted monitoring project.

SCENARIO VI.9

AIRLINE EXECUTIVE: AUTOMATICALLY MONITORING THE PERFORMANCE OF AIRLINE TICKET AGENTS

Andrew Oldenquist

It is proper and necessary that executives monitor worker efficiency. The computer program for doing this is more objective and fewer supervisors are needed, we are told. Job performance, including how quickly and accurately agents produce tickets, is obviously relevant to raises and advancement; hence there is nothing wrong with using this data for that purpose. But the implication is made, later in the scenario, that only this quantitative data is used for evaluation; this would not be an optimal evaluation method, but it is not obviously unethical if workers are told in advance that output is the only criterion (as in piece work or commission work).

It is right to weigh speed of ticket production heavily: The agents are working with lines of people most of whom have reservations, many of whom have time, seating, and connections problems, and all of whom are grateful to get their tickets and boarding passes as quickly as possible.

The agents should be told the methods by which they are evaluated. It is a fault of the executive not to tell them. Being told the method of evaluation is part of treating workers with respect; also it usually is necessary if workers are to know in detail what is expected of them.

The scenario says the agents realized their work was in some way being monitored, and as a result they became stressed, unfriendly, etc. From here on I find the scenario unclear and lacking in verisimilitude. Did they become alienated at the very idea of evaluation? If so, supervisors would alienate them too, and they are unfit to do any supervised work. Or was it just the fact of an unknown method of evaluation, plus a fear that it was mechanical, that alienated them? Are "job performance reviews" conducted with the agents, or without their knowledge?

The ticket agents have only two legitimate complaints, the first, clear and minor (and no justification for alienated and unfriendly behavior) is management gathering evaluation data by a certain method and not telling them. The second, more important but unclear, is that only statistical measurements are used in evaluation and personal needs are ignored. This is a legitimate and ancient ground for complaint, but it has nothing to do with computers and is a seeming gratuitous addition to the scenario.

Their stressed, alienated, inefficient, and unfriendly service appears to be disproportionate to the cause, and certainly unjustified — they are wrong to act that way simply because they suspect their company is monitoring their work by some means. Yet, the reasons are not clear. If nobody paid any attention to special needs, and all evaluation was quantitative and mechanical, there would be strong grounds for complaint. Are we to suppose that management ignores illness, bereavement, machine breakdown, and especially difficult crowds? We are not told, and it would be most implausible.

In any case, the scenario suggests that the agents became alienated because they realized their work was "in some way being monitored," which is no justification for poor work, and not because the evaluation was purely statistical. One would expect, in a true-to-life situation, the electronically gathered performance data to be combined with consideration of an agent's special circumstances, problems, etc.

SCENARIO VI.10

HUMAN RESOURCES INVESTIGATOR: SEARCHING EMPLOYEE RECORDS TO DETECT DRUG USE

David Burnham

Did the company inform the employees about the broad investigation to uncover suspected drug use? Did the investigator who developed the sophisticated monitoring system insist that the company inform the employees? How do we know that the system devised by the investigator is sufficiently accurate to trigger the implicitly serious business of requiring a urinalysis? Is the whole investigation directly related to job performance?

The IRS recently conducted an experiment in which it purchased a computerized marketing list to help it identify individuals who did not file an income tax return. The list had been developed by merging Census tract information with computerized telephone lists and state motor vehicle data. The result was a list of individuals living in each census tract with an estimate of their income, family size, etc. (If they own a Cadillac, it's high, and if they drive a Pinto, it's low.) The IRS tried to match this list against another list of taxpayers. Anyone who appeared on the marketing list but not the list of taxpayers was subject to an official inquiry and possibly an audit. The marketing companies unsuccessfully argued that their lists were so inherently inaccurate that they should not be used for quasi law enforcement purposes. The national association of marketing companies, in fact, said the proposed IRS use of the marketing list violated the association's ethical standards. It requested that none of the marketing companies sell their lists to the IRS. One of the companies ignored this request and the IRS conducted the experiment.

One additional thought. It seems to me that most of the scenarios are based on the question of whether an individual official or computer scientist had lost his or her way. It seems to me that many of the problems go beyond individual responsibility, that the computer has allowed, even required, the development of institutions that are inherently unethical or at least amoral.

SCENARIO VI.10

HUMAN RESOURCES INVESTIGATOR: SEARCHING EMPLOYEE RECORDS TO DETECT DRUG USE

Andrew Oldenquist

First, it is clear that the company had a serious problem and a right to discover which employees used drugs on the job. Second, the new method of discovery used by the investigator is as ethical as the "conventional," since all the records belong to the company and they have a perfect right to look at them. The only question would be the company's right to see the results of the medical exams; but the implication is that the employees waived the confidentiality of these medical records because they voluntarily took the exam, presumably knew its purpose, or at least knew that their company organized the medical exams for the purpose of seeing the results.

The company actually was quite cautious and reasonable in following up positive indications: They could have insisted on an immediate urinalysis, but instead gave them a month to become clean, or leave, or test negative. There can be no right of an employee to privacy about whether he used drugs on the job, given that such use harmed product quality, and was illegal.

The only question I might have is that the investigator should have told the company what his new procedure was; but perhaps he wanted to surprise them with fine results, or thought them too busy to want to see him until he had results.

Someone might argue that every employee, or none, should submit to the urinalysis. But why make unsuspected people take it? One might object if being under suspicion became public knowledge, and embarrassing, at least to the innocent. But we are not told it became public. Presumably they received private letters saying they would have to take a urinalysis; we are not told if it could be taken discreetly, but that could be arranged.

I find it difficult to see any grounds at all for arguing that either the company or the investigator acted unethically.

SCENARIO VI.10

HUMAN RESOURCES INVESTIGATOR: SEARCHING EMPLOYEE RECORDS TO DETECT DRUG USE

Danielle Pouliot

A.2 The investigator is abusive and the comparison he makes with personal data on the employees goes against every principle of professional ethics.

A.3 The company considers the identified employees' as guilty even before proving their addiction to drugs at the expense of their reputation.

B. The firm's request to identify employees suspected of using drugs is contrary to every principle of professional ethics. This practice is similar to a witch hunt since it is very dangerous by giving way to all sorts of abuse, that is to say, every employee who is not productive enough, to the eyes of the employer, may be potentially identified as a drug addict, to his harm. Secondly, if there really are, in the company, persons addicted in such a way to drugs that their productivity is critically low, I do not understand why the company has to investigate so deeply to identify them. Their productivity should be so mediocre that the immediate superior would notice it or complain about it. In the final analysis, we consider that drug addicts have many problems and the way the company identified them does not reflect a willingness to help them. Naturally, things were not the way the company intended them to be, but the company did not assume its responsibility by letting the Human Resources Manager investigate the way he did. In this sense, it is clear to us that the company did not respect the basic principles of professional ethics by not being able to assume its responsibility in managing the problem of drug addiction. Furthermore, the Human Resources Manager neither assumed his responsibilities nor respected basic principles of integrity. The way he identified employees addicted to drugs was dishonest.